A Field Guide to
Mesozoic Birds
and other Winged Dinosaurs

Matthew P. Martyniuk

Vernon, NJ: Pan Aves
2012

Published by Pan Aves
Vernon, New Jersey
www.panaves.com

ISBN-13: 978-0-9885965-0-4

To Rashida,

For always bugging me to add more color,

Carotenoids be damned.

Contents

On the cover: Boluochia zhengi
Title page: Confuciusornis sanctus

Acknowledgments

This book would not have been made possible without huge amounts of data, reference material, support and inspiration. I especially need to thank:

John Conway, for his part in initializing this project, and for the inspiration to broaden its scope to cover all Mesozoic birds.

Mickey Mortimer, whose Theropod Database provided invaluable summaries of available data.

Scott Hartman, Jaime Headden, Ville Sinkkonen, and other artists whose detailed skeletal restorations provided wonderful reference material for some of my restorations.

Roger Torey Peterson, whose Field Guide and Identification System provided the inspiration for the format of this and many other guides to bird identification.

All illustrations created using Adobe Photoshop CS5 and Adobe Illustrator CS5 with WACOM Graphire 3 and WACOM Bambroo Capture tablets. Layout produced using Adobe InDesign CS5. All illustrations and diagrams by Matthew Martyniuk. Human silhouette used in scale charts adapted from Pioneer plaques created by NASA, and which are public domain.

Introduction

Defining "Birds"

Prehistoric birds are often overshadowed in the public consciousness by their larger dinosaurian relatives; *Triceratops*, *Tyrannosaurus*, and decidedly non-bird-like (fictional) portrayals of "raptors" dominate the prehistoric landscape in popular culture from film to TV to video games. Ask even enthusiastic young dinosaur fans to name a few prehistoric birds, and most will be stumped after the obvious choice: *Archaeopteryx*, the famous "first bird" (or "urvogel", an increasingly popular term for the Bavarian fossil species). However, we now know that Mesozoic birds were incredibly diverse, and discoveries since the late 1990s have shown that some of the most popular and well-known dinosaurs, including the "raptors", may be included among them.

The first *Archaeopteryx* fossil, and in fact the first Mesozoic bird fossil ever found under just about any definition of the term (see below), was described by Hermann von Meyer in 1861. Consisting of a single fossil feather unearthed in the Solnhofen limestone quarries of Bavaria, this historic find is now housed in the collections of the Humboldt Museum in Berlin. Complete skeletons soon followed, and revealed a species much different than modern birds. *Archaeopteryx lithographica* ("lithographic ancient wing", referring to the use of the limestone it was preserved in for lithographic printing) had a long tail like a reptile, no beak, numerous small teeth, and clawed fingers, among other primitive features. Subsequent discoveries of small dinosaurs would show striking similarities to *Archaeopteryx*, leading to the hypothesis that dinosaurs and birds were close relatives. This hypothesis would be strengthened with the discovery of species like *Deinonychus antirrhopus*, which were so similar to *Archaeopteryx* that some scientists placed them in the same biological "family". Today, a vast majority of scientists hold the opinion that birds are the dinosaur's direct descendants, and therefore are considered to be a sub-group within *Dinosauria*. Birds are as much dinosaurs as the long-necked elephantine sauropods were: neither group were considered dinosaurs under the original definition of the term, but have been included thanks to later discoveries.

Pretenders to *Archaeopteryx's* title of "first bird" have come and gone over the past several decades. Most infamous of these is the still-contentious *Protoavis texensis* specimen, which is likely to be a chimera made up of remains from various reptiles including dinosaurs and prolacertiforms. Additionally, several sets of very bird-like fossil footprints from the late Triassic and early Jurassic have been reported, complete with reversed halluces (the first toe of the foot which opposes the others in perching). However, there is some doubt about whether those truly come from birds and not more primitive theropods or even bipedal reptiles related to crocodiles, and it is possible that the age of some tracks has been misinterpreted. Despite the dubious nature of these challengers, *Archaeopteryx* has retained its title largely because of the nature of classification itself. Historically, the term "bird" has become strongly associated with the group *Aves*, named by Carolus Linnaeus (widely regarded as the father of biological classification) in 1758. Linnaeus designated Class Aves (which means "birds" in Latin) as one of the major divisions of life, ranked highly in his famous system of taxonomy (the familiar hierarchy of Kingdom, Phylum, Class, Order, Family, Genus, and Species), lesser in status to mammals (Class Mammalia) but greater than Reptiles (Class Reptilia) among the animals with backbones (Phylum Chordata).

After centuries of use, Linnaeus' system of classification began to show its age. By the 1960s, biologists had become increasingly dissatisfied with various incompatibilities between the Linnaean system and evolutionary theory. Linnaeus had, after all, devised his system before Darwin demonstrated the common descent of life. Over the next several decades, many scientists, especially those working in the field of vertebrate paleontology, would largely abandon the Linnaean system in favor of one based on clades (groups containing certain species, their common evolutionary ancestor, and all other descendants of that ancestor, no matter how modified from the original form).

The use of clades in place of ranked classes addressed another shortcoming of the Linnaean system: its lack of concrete definitions. Linnaean classes were determined by a vague set of characteristics (diagnoses) and had no set definitions, making their use often rather subjective. Clades, on the other hand, are required to have strict definitions, and are defined not by characteristics themselves, but by the evolutionary relationships revealed by rigorous analysis of those characteristics. In phylogenetic naming, *Mammalia* doesn't mean warm-blooded vertebrates with hair that feed their young with milk, but rather the group of vertebrates that contains placentals, monotremes, and marsupials, their common ancestor, and all other descendants of that ancestor, either known now or to be discovered later.

Applying this method to birds has proven divisive. Originally, scientists like Jaques Gauthier (one of the fathers of phylogenetic naming) defined *Aves* the way *Mammalia* came to be defined--as a "crown group", that is, the group containing all modern bird groups and any species closer to them than to more distantly related prehistoric lineages. This use of *Aves* preserves Linnaeus' usage (he didn't know of any prehistoric birds), but also excludes fossil bird groups such as *Archaeopteryx* and even those very similar to modern birds like *Ichthyornis*. Subsequently, attempts were made to re-define *Aves* to include *Archaeopteryx* for historical reasons; after all, it has always been considered the "first bird". Critics called these attempts arbitrary: why include *Archaeopteryx* and not their closest relatives (such as *Microraptor*), which were more like modern birds in some respects? As of this writing, a body known as the International Society for Phylogenetic Nomenclature is being organized to govern the naming of clades, and it seems likely that it will adopt the crown group usage as official.

When this usage is officially adopted, it will leave most Mesozoic birds outside of *Aves*. This will only require a semantic change; true "birds" will be restricted to the group of modern-style birds only. Non-avian members of the lineage leading to modern birds such as *Ichthyornis* will be considered "stem-birds" (and, somewhat ironically, also a "non-avian dinosaur" genus, as all birds are included in the clade *Dinosauria*).

So what is a "bird"? As a vernacular rather than scientific term, this is a matter of loose convention rather than scientific precision. Most common definitions of the word in English hinge on certain key aspects: egg-laying, feathers, and flight. If this usage is adopted, *Archaeopteryx* may still be considered birds, depending on their controversial flight abilities (though they were almost certainly at least glissant, i.e. capable of passive gliding). *Microraptor* are probably "birds" as well under this definition. Of course, not all birds fly, and just as there are ground birds and flightless birds today, many Mesozoic lineages lost flight, or evolved before flight. Therefore, any dinosaurs which would now be included by one of the first definitions proposed for the clade *Aves*, by Allan Charig in 1985, are featured in this book. Charig's definition of *Aves* linked it to the presence of feathers. While he did not specify what he meant by feathers, for the purposes of this book I am using a conservative interpretation: feathers with a fully modern anatomy, consisting of a rachis (central "quill"), and a vane comprised of barbs linked together by barbules (see diagram on p. 39). This is also the same characteristic that was used to classify *Archaeopteryx* as birds in the first place, so it seems most appropriate for use in this work. This definition is equivalent to the modern clade *Aviremigia*, that is, all winged dinosaurs.

By their nature, feathers do not fossilize well, nor often, so there may certainly be some species or even entire lineages which were "birds" under the above definition, but are excluded here because we do not have enough information about their plumage. While some primitive coelurosaurians, such as compsognathids and deinodontoids, had "feathers", these appear to consist entirely of filamentous and/or down feathers, not vaned feathers. The plumage of ornithomimosaurs is also known to be downy, and while adults appear to have had feathers with central quills on at least their arms, the exact anatomy of these structures is currently unknown. Among maniraptorans, one specimen each of alvarezsaurians and segnosaurians preserve feathers, but both lack evidence that they were vaned (possibly due only to poor quality preservation). One intriguing species known only from feathers, *Praeornis sharovi*, had rachides (quills) with barbs, but lacked true barbules, having only strange ridges of tissue lining the individual barbs (possibly an evolutionary precursor to barbules), and so is the closest outlier to the definition of "feather" used here. All of those species are excluded from this book, though future discoveries may reveal that they had modern feathers after all.

Below: The evolution of modern birds from feathered dinosaurs.
From left to right: Dilong paradoxus (Deinodontoidea), Nqwebasaurus thawzi (Ornithomimosauria), Haplocheirus sollers (Alvarezsauria), Yixianosaurus longimanus, Xiaotingia zhengi (Deinonychosauria), Archaeopteryx lithographica, Confuciusornis sanctus (Confuciusornithiformes), Bohaiornis guoi (Enantiornithes), Apsaravis ukhaana, Ichthyornis anceps

Bird Origins & Evolution

Bird Ancestry

Since the 1970s, a consensus has emerged among scientists that birds evolved directly from a certain branch of specialized bipedal, carnivorous or omnivorous dinosaurs, the *Coelurosauria*. In fact, the anatomical thread of transition from a lizard-like early diapsid reptile to a modern bird can be followed throughout the entire dinosaurian lineage. While a few dissenters to this theory remain, they have so far failed to propose alternative models of bird origins and to support them with rigorous, repeatable studies.

All researchers agree that the closest living relatives of birds are crocodylians; this view has been overwhelmingly supported by both fossil evidence and genetic evidence showing crocodiles to be more closely related to birds than to lizards. The last shared common ancestor of birds and crocodylians (i.e. the ancestral archosaur) probably lived during the early Triassic period at the dawn of the Mesozoic era. Shortly thereafter, the archosaurian lineage split into the crocodile line (*Pseudosuchia*, the inappropriately named "false crocodiles") on one hand, and the bird line (*Ornithosuchia*, or "bird crocodiles") on the other. While it's difficult to imagine two types of animals more different in terms of anatomy, the ancestor of birds and crocodiles did share some important characteristics of both. For example, both birds and crocodiles have four-chambered hearts, and so it is likely that their Triassic common ancestor did as well. This common ancestor, or "concestor", also probably built nests out of vegetation rather than burying eggs in soil, and exhibited some degree of parental care--even modern crocodiles will look after and protect their young from predators for a short time after they hatch. However, this concestor was still more crocodile-like in appearance than bird-like. Though the sprawling, dragging gait of modern crocodylians is partly a result of their aquatic specializations, the bird/crocodile concestor was still probably a sprawling quadruped.

The next major branching event in the bird lineage was the split between the ancestor of birds and pterosaurs, the highly diverse flying vertebrates of the Mesozoic which include the pteordactyls. This concestor, of the group *Ornithodira*, was likely also a semi-sprawling quadruped, and still laid eggs with leathery shells rather than the rigid, fragile eggshells of birds. Interestingly, all known pterosaurs were covered in dense coats of fur-like filaments called pycnofibres. It is possible that these were present in the bird/pteorsaur concestor and represent the earliest stages (Stage I, described below) of feather evolution. However, we will need a more complete record of skin coverings in ornithodirans before we can test this idea.

Around the base of the ornithodiran lineage and that of its immediate sub-group *Dinosauromorpha*, the ancestors of birds began to walk upright. This probably began as a mode of fast locomotion, similar to the way crocodylians adopt a more upright stance when walking quickly. The early ornithodirans, while still primarily quadrupeds, evolved very long hind legs, and were probably bipedal runners. In the slightly more advanced *Dinosauriformes*, all four limbs were probably held beneath the body, and by the advent of *Dinosauria* proper, a specialized hole had opened in the hip socket (a perforate acetabulum), locking the hind legs into an upright stance permanently, making sprawling the hind limbs impossible for all dinosaurs, except those advanced species which modified this arrangement for swimming and climbing. The first dinosaur, the concestor of birds and ornithischians (another highly inappropriate and misleading name meaning "bird hips"), was probably an omnivorous biped with five fingers (three of which bore claws) and five toes. It likely existed in the early part of the Late Triassic.

Soon after the split between the ancestors of birds and ornithischians, the eusaurischian concestor (the common ancestor of birds and sauropod dinosaurs) evolved the beginnings of the avian lung system, which involves extensions of the lung in the form of air sacs invading many of the bones in the skeleton, making them 'hollow' and air-filled. With the advent of the group *Theropoda*, bird ancestors became more fully dedicated to a bipedal, running lifestyle, and shortly thereafter the foot became modified to have three primary toes, with a fourth toe small and high on the ankle, like a dewclaw, and the fifth toe was lost completely. That small dewclaw, or hallux, would be retained and later enlarged and modified in several theropod groups to assist in walking, in prey capture, and in perching. The two claw-free fingers of the hand were slower to be lost, but by the time the group *Avetheropoda* appeared, only three fingers remained.

At some point, probably after the avetheropods had evolved, members of the bird lineage developed one of their most important characteristics: feathers. The first feathers were probably simple, hair-

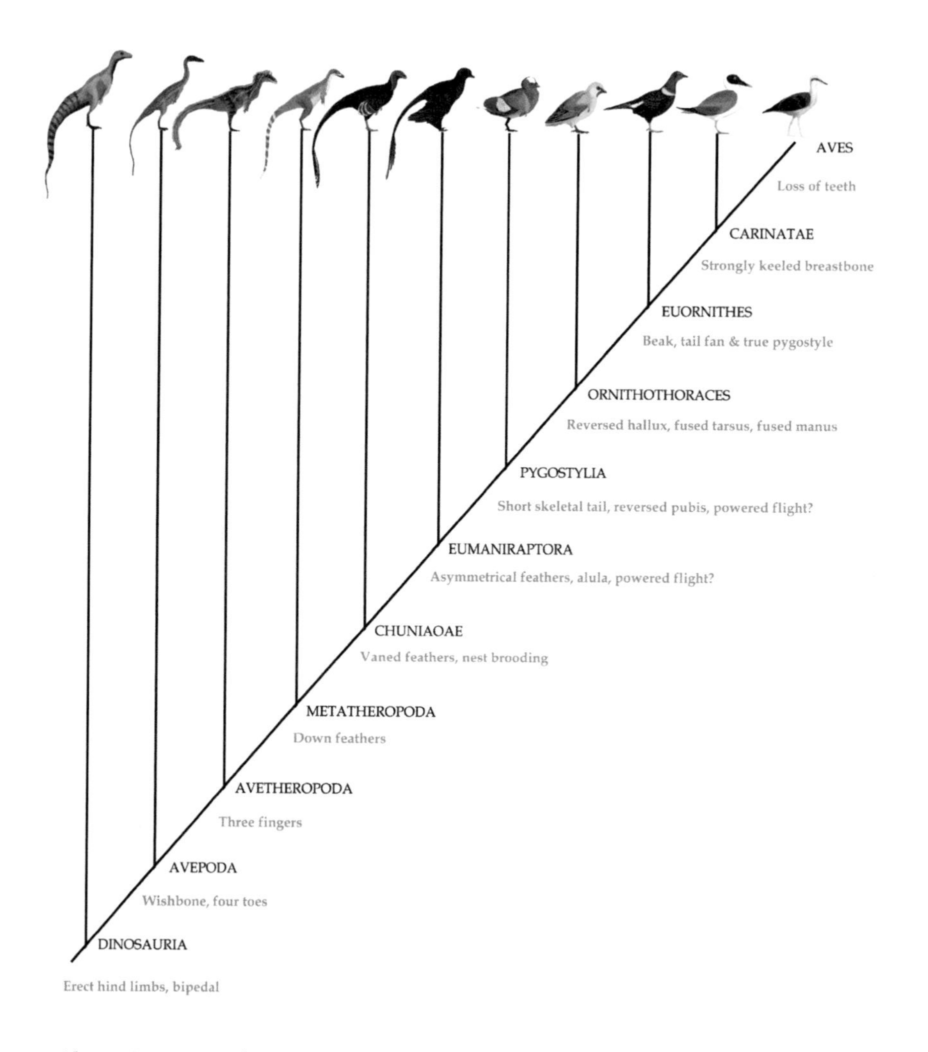

Above: Sequence of aquisition of major avian characteristics within Dinosauria.

like filaments made of beta-keratin, and the earliest examples have been tentatively identified in the possible primitive coelurosaurian *Sciurumimus* (that is, if these aren't the same structures seen in pterosaurs and/ or some ornithischians). The earliest down feathers appear in the group *Tyrannoraptora*, in both primitive members of the group (the deinodontoids and compsognathids), so it's likely that their actual origin is somewhat earlier than their Middle Jurassic concestor. Evidence from specimens of the ornithomimosaurs *Dromiceiomimus*, a group which probably branched after compsognathids but before maniraptorans in the Middle Jurassic, suggests that these may have had pennaceous, if not closed-vaned, feathers on the arm forming the earliest wings. The slightly more advanced group *Segnosauria* also preserves evidence of relatively long wing feathers, but these appear filamentous. More evidence and better-preserved specimens of these groups will be needed before the exact nature of their feathers can be determined.

Truly modern, definitively vaned feathers appeared soon afterwards, in the Middle Jurassic concestor of birds and caenagnathiformes. This same group (*Chuniaoae*) is where nest brooding first appears in the fossil record, probably due to the superb ability of vaned feathers to regulate heat. Aerodynamic, assymmetrical feathers appeared next, in the group *Eumaniraptora*, along with vaned feathers anchored to the minor digit of the hand, forming the first rudimentary alula ("bastard wing"). These adaptations strongly suggest that by this point birds had at least begun taking the first steps toward flight.

Further refinements to bird anatomy also seem to have been biased towards increasing aerodynamic ability. The loss of the long tail, for instance, occurred prior to the evolution of the group *Pygostylia*, and from there flight-related adaptations exploded. In the *Ornithothoraces*, the shoulder anatomy was modified allowing the wings to be extended vertically, resulting in a full flight stroke, and the major and minor digits of the hand fused together, as did the bones of the tarsus. The first toe, which had gone from a dewclaw to a prey capture and climbing aid in eumaniraptorans, evolved into the reversed hallux in ornithothoraceans, allowing birds to perch on small branches.

In *Euornithes*, the first avian beaks evolved alongside the toothy jaws of these birds' ancestors, and the tail feathers evolved into an expandable fan-like structure anchored to a true pygostyle. Further refinements to the wing anatomy allowed launching into the air from the

Above: Juravenator starki, *an early coelurosaur covered in both scales and simple (Stage I or II) feathers.*

ground or water, negating the need to climb up trees, and the wing claws were greatly reduced by the level of *Carinatae* (though small claws are still present in many modern birds). While the keel on the underside of the breastbone first appeared in *Ornithothoraces*, it became greatly enlarged in carinatans, signalling the advent of modern flight driven primarily by strong breast muscles. Finally, in the concestor of all modern birds, the teeth were lost, resulting in the characteristic toothless bill of *Aves*.

The Origin of Feathers

The details regarding exactly how feathers first evolved have been historically contentious and are still not entirely clear. However, hypotheses based on the developmental stages of modern feathers (such as those proposed by Richard Prum, e.g. Prum 1999) seem to match the fossil record fairly well at this time. According to Prum, the first feathers (Stage I of feather evolution) would have been simple quill-like structures emerging from a follicle in the skin, similar to the way scales and hairs emerge from skin follicles. These simple quills, often referred to as "protofeathers", would have been monofilaments with no branching structure, equivalent to the rachides that form the central structure of modern feathers. The presence of Stage I feathers in the fossil record is uncertain. The earliest probable examples come from the Late Jurassic coelurosaurian or megalosaurian genus *Sciurumimus*. The long, somewhat flattened "elongate broad filamentous feathers" (EBFFs) of some primitive feathered dinosaurs including the segnosaurian species *Beipiaosaurus inexpectus* may represent either Stage I feathers, novel structures derived from more advanced feather types, or simply vaned feathers distorted by crushing. Similarly, the unusual monofilament quills of ceratopsians *Psittacosaurus* and heterodontosaurians *Tianyulong* may represent Stage I feathers. If this is the case, the first feathers must have evolved at or before the origin of *Dinosauria*. It is even possible that "pycnofibres", down-like filaments found in pterosaurs, evolved from Stage I feathers. If this is the case, feathers may have their origins in the early Triassic period at the origin of the group *Ornithosuchia*, shortly after the ancestors of birds split from the ancestors of crocodiles.

Stage II feathers evolved when the feather follicle collar (equiva-

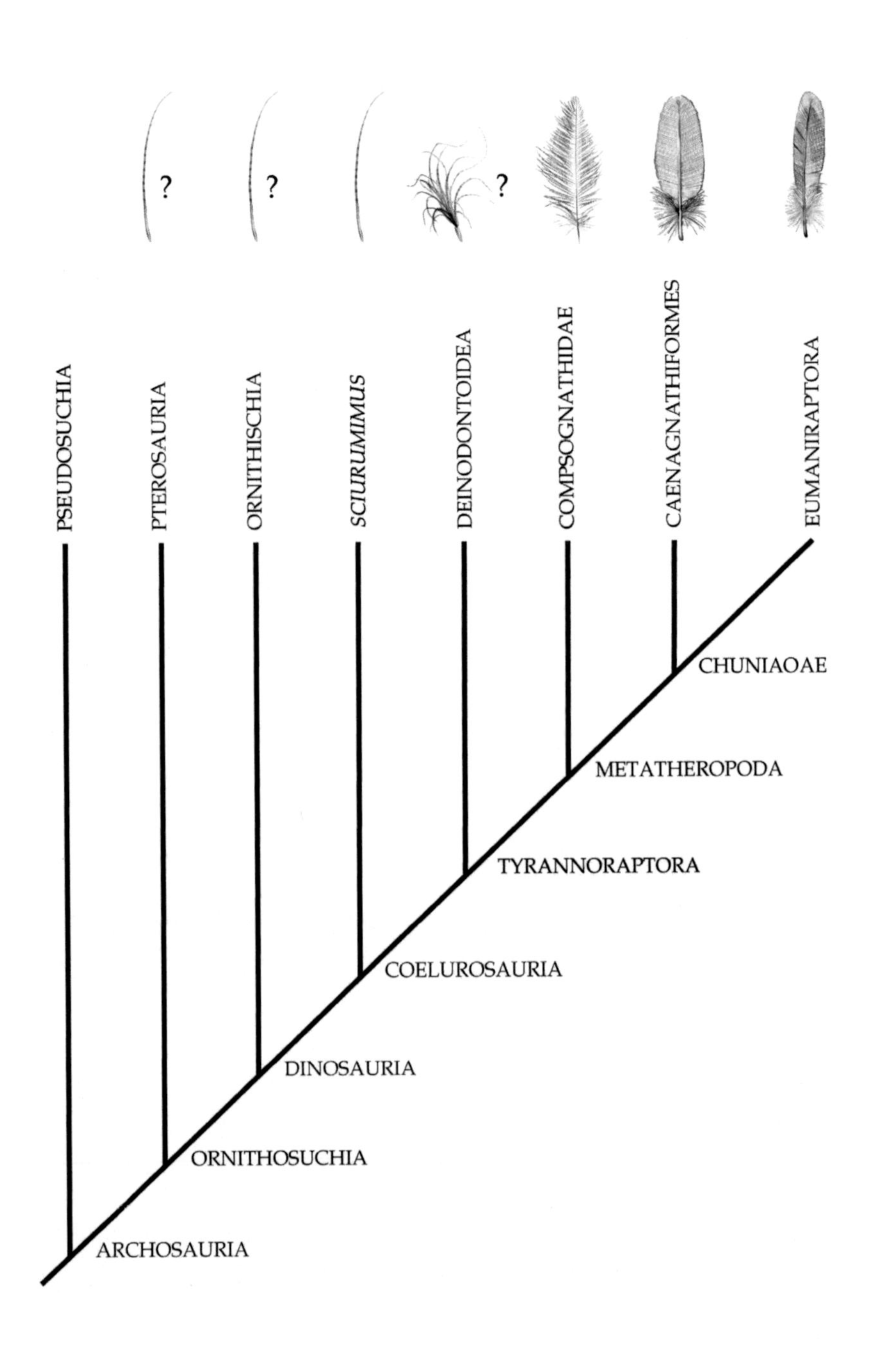

Above: Approximate sequence of feather evolution mapped to the dinosaurian family tree. Feather types, from left to right: Stage I, Stage II, Stage III, Stage IV, Stage V.

Above: Sinosauropteryx prima, *a compsognathid exhibiting Stage III down-like feathers.*

lent to the calamus of modern feathers, used as the nib in quill pens) gave rise to several filaments instead of forming a single long quill. This type of feather still exists today in the form of down feathers, the simplest form of which is made up of a short calamus and numerous long, soft filaments that form a tuft rather than a vane. All of the filaments (called barbs) in a down feather are anchored to the central calamus rather than to a central quill or rachis. Down feathers have been identified in compsognathids like *Sinosauropteryx prima*. More primitive theropods, like *Dilong* and *Yutyrannus*, may have had Stage II down feathers, but the preservation in relevant fossils is too poor to be certain. At the very least, we can assume that down feathers emerged at or near the base of the advanced theropod group *Coelurosauria*, and are present in all more advanced theropods including modern birds.

The next stage in feather evolution according to Prum is uncertain. Stage III could have been basic down with the addition of smaller, microscopic branching structures on their barbs (called barbules). Alternately, the next stage may have involved the barbs beginning to grow in a helical fashion up along a central filament, or rachis, in a type of advanced down known as a semiplume. Either barbules or a semiplume structure could have evolved first, or they could have evolved simultaneously. The fossil evidence is uninformative on this point, as neither barbuled down feathers nor semiplumes have been definitively identified in fossils of non-aviremigians. However, some potential fossil feathers, classified in the species *Praeornis sharovi*, appear to show a central rachis and thick barbs lacking differentiated barbules, instead showing solid ridges on the barbs. This may be an early form of, or derivation of, an unbarbed semiplume. However, the identification of the *Praeornis* feathers is controversial and some researchers have even proposed that they are not feathers at all, but cycad leaves, though chemical testing seems to indicate that they are indeed animal in origin.

Stage III feathers have been positively identified in the compsognathid *Sinosauropteryx prima*. This species appears to have had both moderately long rachides with long barbs (lacking barbules) along their lengths. Stage III feathers may also have been present in ornithomimosaurians like *Dromiceiomimus brevitertius*, in which juveniles appear to have been covered in Stage II or III down and adults appear to have had primitive wing-like arms with anchor points for pennaceous remiges and coverts, though whether or not these were vaned (making them Stage IV) is currently unknown (Zelenitsky & al. 2012).

Stage IV represents true, modern bird feathers. The primary barbs and simple secondary barbules of the semiplume evolved another level of branching, in the form of tertiary hooklets. These allow the barbules to hook together and link barbs in a single, closed vane. Stage IV feathers are first seen in the caenagnathiformes and basal eumaniraptorans. Vaned feathers are aerodynamic and useful for flight, especially in some more advanced derivations with asymmetrical vanes (where the barbs are longer on one side of the quill than the other). Prum called asymmetrical vaned feathers Stage V. Stage V, or flight feathers, are first found in ornithodesmids and other eumaniraptorans, including *Archaeopteryx*.

The diagram on page 20 illustrates the approximate appearance of the various feather stages in the archosaurian family tree. However, it should be noted that new feather types often evolve from these basic structures, and in many cases advanced feather types are lost in flightless groups. For example, among modern ratites (ostriches, emu, kiwi, etc.), vaned feathers have been lost. The remiges and rectrices of these birds have effectively returned to Stage III semiplumes with central rachides but lacking barbules. This is also true of the known feathers of hesperornitheans. In more recently flightless birds, like the Kakapo (which still retains some gliding ability), the wing feathers have reverted from Stage V to Stage IV, with symmetrical vanes similar to caudipterids or *Anchiornis*. Given enough time, it is likely that these feathers would lose their barbules entirely as gliding becomes less and less an important part

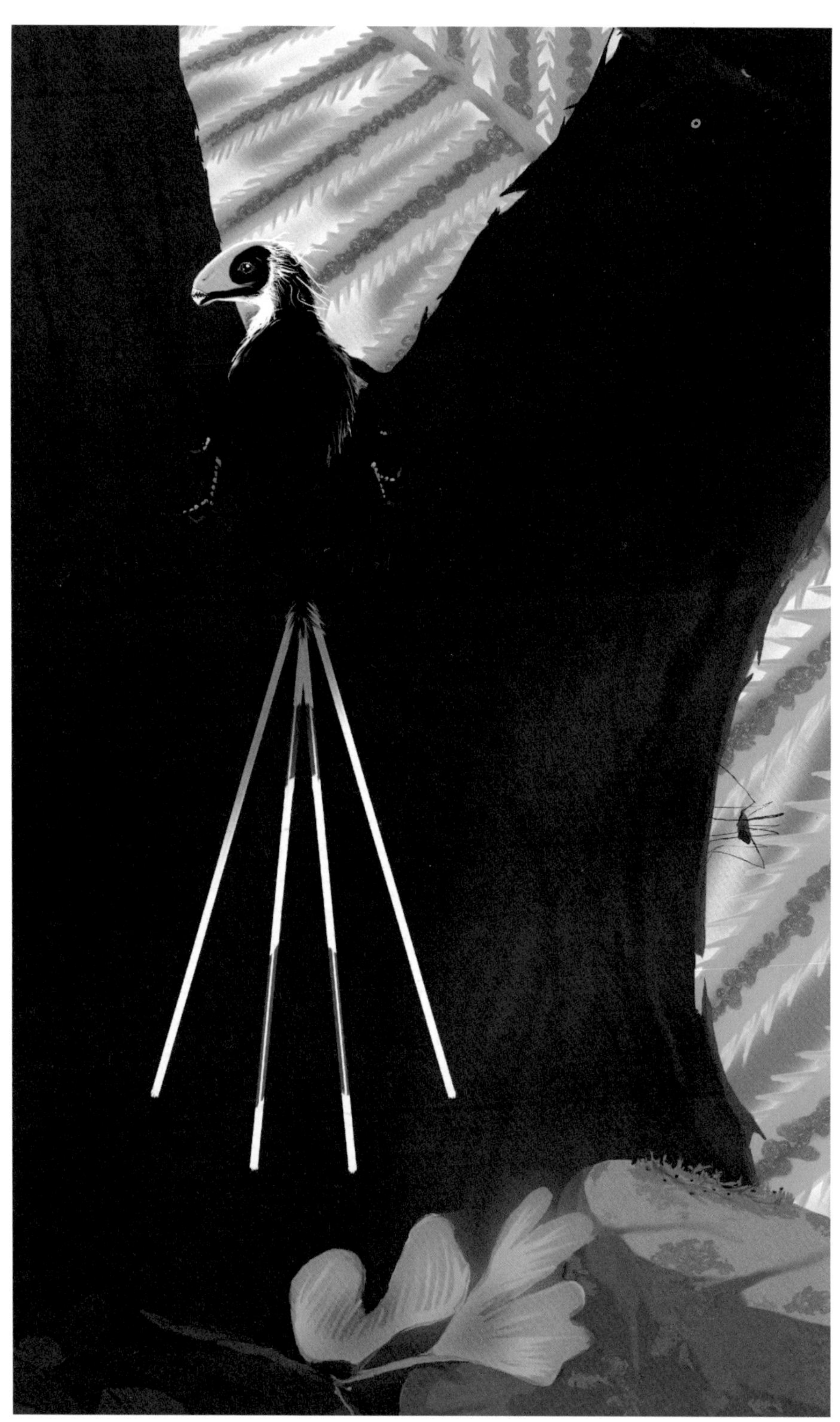

Above: Illustration of an Epidexipteryx hui, *part of the Dauhugou Fauna.*

of the bird's lifestyle. This degeneration of the remiges appears to be a general trend in birds as they lose flight, probably because vaned feathers are more costly to maintain than plumulaceous or semiplume feathers.

Novel feather types that have evolved from the feather stages listed above include the EBFFs of segnosaurians, comprised of long, flat, monofilament quills; the ribbon-tail feathers of the so-called "opposite birds" (enantiornitheans), likely produced by extending and flattening the rachis and reducing the vane for most of its length; filoplumes, small feathers with the barbs restricted to the tip of a thin rachis, which may act as sensory organs; bristles, the opposite of filoplumes, with barbs at the base but whisker-like naked rachides at the tip; powder down, plumulaceous feathers which never molt but continuously grow and flake at the barb tips to create a powder used in preening and waterproofing; and display feathers, like those of some birds-of-paradise, in which the barbs have fused for most of the rachis length creating a solid sheet. While most of these are known to be present only in modern birds, some have evolved more than once, and it is possible that some Mesozoic birds developed similar structures which have yet to be found in the fossil record.

The First Birds

Since the initial discovery of the species in 1861, *Archaeopteryx lithographica* had been considered the first (i.e. earliest known) bird, classified as such based on the presence of true feathers. For decades, these remained not only the earliest, but the only Mesozoic fossils known to possess the impressions of feathers (an early specimen of *Parahesperornis* also preserved feather traces, but this was to be expected of a more "advanced" group). By the time more Mesozoic-age feathered fossils were discovered in the 1990s, cladistics had all but replaced Linnaean classification in paleontology. Therefore, when more species with feathers and wings were found, they were not considered birds, but rather feathered, non-avialan dinosaurs. Dinosaur species like *Sinornithosaurus millenii* and *Microraptor zhaoianus*, despite possessing feathered wings and possibly some form of flight, were not considered birds because they were thought to fall just outside the clade containing modern birds and *Archaeopteryx*. Either way, these fossils hailed from the Yixian and Juifotang formations of Cretaceous China, later than the Jurassic *Archaeopteryx*. Birds or not, *Archaeopteryx* were still the earliest animals with true feathers.

The discovery of feathered dinosaurs in the Chinese Daohugou Beds muddied the waters. To date, three feathered species are known from the Daohugou: *Pedopenna dauhugouensis*, *Scansoriopteryx heilmanni*, and *Epidexipteryx hui*. The Daohugou dates to earlier than the Yixian, but how much earlier is controversial. The average age found in published scientific literature is Late Jurassic, contemporary with or slightly earlier than *Archaeopteryx*. Some studies suggest they are Middle Jurassic, which would make them solidly earlier than *Archaeopteryx*.

The Daohugou species ceded their titles as the first feathered animals in 2009, when feathered fossils were described from the even earlier Chinese Tiaojishan Formation, dating to between 161 and 155 million years ago, at least five million years older than *Archaeopteryx*. The first of these to be announced to the public were *Anchiornis huxleyi* and *Xiaotingia zhengi*, both closely related to *Archaeopteryx*. Like the Daohugou *Pedopenna* and the Jiufotang *Microraptor*, *Anchiornis* possessed not only well-developed wings, but wing-like feather structures on their hind legs, strongly suggesting that birds ultimately descended from gliding, four-winged forms. As of this writing, *Anchiornis huxleyi* and *Xiaotingia zhengi* are together the first "birds" in the broad sense of the term. As noted above, many modern researchers restrict *Aves* to the crown group, and in this sense the earliest known true birds are actually significantly later than *Archaeopteryx*. The earliest known true avian bird that can be referred to as such with a degree of confidence is *Austinornis lentus* from the Late Cretaceous of Texas, 85 million years ago. Molecular studies suggest, however, that true birds originated closer to 130 million years ago (about 20 Ma after *Archaeopteryx*), and at least one potential avian is known from that same timeframe in France: *Gallornis straelani*; though its identification is far from certain.

Mesozoic Bird Diversity

Despite their absence from popular media in favor of larger, scalier dinosaurs, Mesozoic birds were very diverse. As evidenced by early classification schemes, these first "birds" can generally be divided into two types: long-tailed and short-tailed. Pre-cladistic classifications divided these into the subclasses "Sauriurae" ("lizard tails," for their long, bony, reptile-like tails) and "Ornithurae" ("bird tails," which had evolved the modern configuration of an extremely shortened tail). However, these represent evolutionary grades, rather than clades; the short-tailed birds evolved from long-tailed ancestors, and so long-tailed birds as a whole cannot be considered a distinct group.

Chiappe (2006) termed these two basic types "fan-tailed birds" and "frond-tailed birds", due to the distinct configuration of feathers the differing tail lengths produced. In frond-tailed birds (the most primitive type), tail feathers (rectrices) were generally arranged along at least part of the length of the long skeletal tail, producing a shape reminiscent of a fern frond. Each pair of rectrices usually anchored to one of the tail vertebrae. Modern birds are fan-tailed; due to the shortening of the tail, all tail feathers are anchored to a single ploughshare-shaped bone, called the pygostyle, and so form a fan that the bird can often fold and unfold with the help of the attendant musculature of what is known as the rectrical bulb.

However, this distinction is not cut and dry among Mesozoic birds. The advent of the true fan tail seems to have come later than previously thought. Many early short-tailed birds, like the long-winged, beaked confuciusornithids, only had a pair of specialized ribbon-like feathers on the tail, and then only in one sex. More than half of known *Confuciousornis sanctus* specimens lack rectrices; their tails are merely fluffy stumps, covered in contour feathers. The stump-tailed forms may represent females, or alternately, individuals undergoing a molt. The ribbon-like feathers of (presumably) males are also found in many specimens of enantiornitheans (the "opposite birds"), usually in pairs of two or four feathers, but up to eight in some species. All have fused, shortened tail vertebrae at the tail tip, similar to the modern pygostyle. However, in these birds the fused tail structure is simple and rod- or

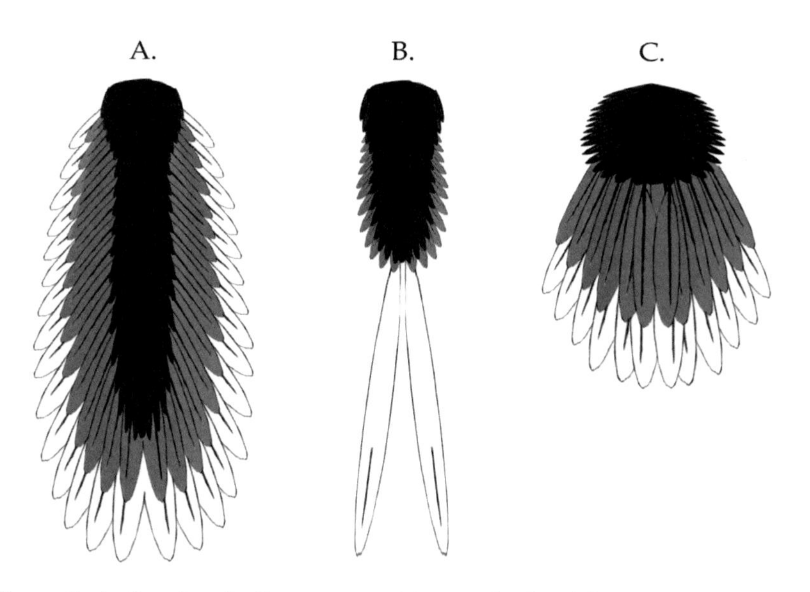

Above: Basic diversity of tail structure in Mesozoic birds. A) frond-tailed (e.g. basal avialans); B) ribbon-tailed (e.g. enantiornithes); C) fan-tailed (euornithes)

dagger-shaped, not the complex pygostyle with attendant musculature of modern birds, and so the tail feathers could not have been opened or closed to aid in landing or turning in mid-flight. These intermediate short-tailed birds are perhaps best termed "ribbon-tailed birds" after this unique configuration.

True fan-tailed birds seem to be more advanced, and consist mainly of modern birds (*Aves*) and their closest relatives (collectively the euornitheans, "true birds"). These are distinguished by the presence of a fan of graduated rectrices, with at least the most posterior two attached directly to a ploughshare-shaped pygostyle. The pygostyle is also associated with a muscular organ called the *bulbi rectricium* (rectrical bulb), to which all the rectrices anchor. The bulb is what allows the feathers to expand and contract in a fan-like manner, allowing for far greater maneuverability than the forebears of these species could achieve. In birds without this structure, the tail feathers would have been largely immobile and in most cases were purely for display.

Frond-tailed birds, ribbon-tailed birds and fan-tailed birds all co-existed during the rapid diversification of avialans in the early Cretaceous. Most flying birds of the late Cretaceous were members of the two primary short-tailed lineages: *Enantiornithes* (ribbon-tailed) and *Euornithes* (fan-tailed). *Enantiornithes* were the dominant flying birds of the Mesozoic, but all became extinct along with most other remaining dinosaur lineages 66 million years ago. Only a few lineages of euornitheans survived, and these were all members of the groups that would eventually produce ratites, fowl, and shorebirds, and which diversified into all of the other modern bird groups during the Cenozoic era.

The Evolution of Flight

Modern birds are perhaps best characterized by their ability to fly, that is, generate lift under their own power. Most bird groups today are comprised of strong fliers, and researchers in the past have often been tempted to conflate "bird" with "flight". However, our current understanding of the fossil record indicates the origin of birds (or at least of modern-style, Stage V feathers) must have predated powered flight.

Historically, there has been a debate over whether flight evolved from the "ground up", that is, from strictly ground-running ancestors, or from the "trees down", from arboreal, climbing ancestors. This dichotomy has often been treated as a false one, as it is entirely possible that some combination of the two ultimately resulted in the flight of modern birds. However, both hypotheses do make testable predictions which can be compared with the (admittedly limited) available fossil record. A strict trees-down origin of flight would predict that bird ancestors were already arboreal creatures that would show adaptations for climbing. It would also be likely under this model that flight first passed through a gliding phase, with animals using their wings to passively extend leaps between branches.

On the other hand, a strict ground-up model would predict flapping flight to emerge first, as running animals used their wings to extend leaps, propel themselves up steep inclines, and ultimately achieve ground-based launches into the air. Furthermore, it might predict that the reversed hallux would only begin to evolve after a complete flight stroke had been achieved, allowing previously terrestrial animals to begin perching in trees (Ohmes 2012).

Unfortunately, the fossil record doesn't seem the be a good fit for either of these scenarios. The most primitive winged dinosaurs, such as *Microraptor* and *Confuciusornis*, had hind limbs apparently devoid of any clear climbing adaptations. The perching foot wasn't perfected until the ornithothoraces, when powered flight had probably already been achieved, but the first stages of modification to the hallux are seen as early as the confuciusornithids, if not the deinonychosaurians. However, these primitive forms also did not have the shoulder anatomy that would have allowed a flight stroke, and may have been incapable of launching from the ground. Even early ornithothoracians may have needed to climb to become airborne.

At the moment, a middle ground hypothesis regarding the origin of flight is the only one favored by this evidence. It is likely that in the ancestors of birds, flight and arboreality actually co-evolved. The arboreal adaptations of bird ancestors seem to have advanced simultaneously with those related to flight. It seems that very rudimentary gliding flight or incline running was utilized in the earliest winged birds, slightly extending the range of these animals into the trees to escape predators or access prey out of their competitors' reach. This can be seen as a hybrid hypothesis, "ground up into the trees." Once birds had become fully arboreal, flight could be perfected, but the initial stages of flight evolved hand-in-hand with the initial stages of arboreality.

In modern birds, flight takes a variety of forms, which can be broadly defined in terms of five distinct styles: flapping, flap-gliding, bounding, soaring, and bursting (Close & Rayfield 2012). "Flapping" refers to continuous flapping, a style employed today by birds like ducks and flamingos, which flap almost continuously during flight. Flapping birds come in a variety of sizes, but all tend to have a high wing loading (i.e. small wings relative to the weight of the bird). Flap-gliding is a common form of flight, seen in crows and gulls among others, where flapping is intermittent, interrupted by periods where the wings are extended in a glide. Bounding is another form of intermittent flight, seen mainly in small birds, where instead of gliding, birds enter a "ballistic" phase in between flapping phases. Here, the wings are folded against the body rather than extended, causing the bird's body to be propelled aerodynamically through the air like a bullet. Soaring includes two sub-categories of flight: dynamic and static. In static soaring, birds tend to have low wing loading, broad wings, and slotted wing tips, allowing them to passively exploit rising columns of thermal energy. Static soaring birds include mainly large species like vultures. Dynamic soaring birds exploit wave energy of the air, usually close to the surface of water, and tend to have higher wing loading with long, narrow wings. These include large seabirds like albatross and gannets. Bursting flight is found mainly in ground birds that can only fly for very short distances and cannot sustain long bouts of flapping flight, like quail. These birds employ flight only in short bursts, usually to flee predators (Close & Rayfield 2012).

A 2012 study by Close and Rayfield found a rough correlation between these five flight styles and the shape of the wishbone (furcula), allowing the flight styles of some Mesozoic birds to be deduced. As might be expected, the most primitive birds, such as *Archaeopteryx* and *Confuciusornis*, which had very broad, U-shaped wishbones, were found to group with flightless, soaring, and gliding birds. While this is probably due to the fact that advanced wishbone shape had not yet had a chance to evolve in these primitive birds, it does support the data presented by other studies suggesting that they were only capable of weak flapping, if any, and were probably predominantly gliders.

As expected by their high diversity, enantiornitheans appear to have occupied a wide range of flight styles, including flap-gliding and bounding. However, many enantiornitheans did not group with any living birds in the 2012 study. This is due to the unique shape of their wishbones, which were V-shaped and had a long backward prominence that may have partially taken on the role served by the keeled breastbone in euornitheans. So enantiornitheans may have achieved the standard flight styles in a different way and used different muscles than modern birds, or they may have had a unique style of flight that is now extinct (Close & Rayfield 2012).

These types of studies suggest that flight emerged among the most primitive birds but was not "perfected" (i.e. achieved the flight styles of modern birds) until the euornitheans evolved in the early Cretaceous. Non-euornithean birds were probably poorer fliers (or at least very different sorts of fliers) than modern birds in a number of ways. Non-ornithothoracine birds (those more primitive than enantiornitheans) did not have a full upstroke, due to the situation of the shoulder girdle relatively close to the belly rather than the side or back of the rib cage. This would have limited those birds to gliding or, at best, short and weak bursts of powered flight. Enantiornitheans had likely achieved a full upstroke, but probably did not yet have a modern downstroke where the arm moves outward and forward (protraction), making the flight stroke overall less powerful in terms of generating thrust (Close & Rayfield 2012). As discussed above, enantiornitheans had shortened their tails but consequently lost most of their tail feathers. This severely restricted their ability to control their flight speed and would have made precise landings nearly impossible. The retention of wing claws in these species indicates that they may still have relied on climbing to a large degree when moving around trees, taking off, and landing.

The evolution of the fan tail and rectrical bulb in euornitheans allowed higher maneuverability and speed control, though in many primitive species the fan-tail was long and extravagant, suggesting it may have evolved initially for display, possibly an elaboration of the ribbon-tails of more primitive birds. Euornitheans soon reduced their wing claws, however, suggesting that the need to climb was becoming lessened. This is probably due in large part to the combination of the fan-tail and a more aquatic, shorebird-like lifestyle, as well as improvements in the wing anatomy that led to an increased ability to launch from flat ground or water.

Restoring Mesozoic Birds

While Mesozoic birds and other ancient dinosaurs are long extinct, it is possible to illustrate them within a reasonable margin of error by combining fossil evidence with principles gleaned from observing modern species and good modern analogues. Thanks to a small but growing number of fossil birds with preserved skin and feathers, it is possible to get a general idea of how feathers evolved, how they were distributed in the first birds, and their ancestors, and how feather types and patterns differed between major groups.

It is helpful to keep in mind, however, that plenty of speculation needs to be employed when comprehensively illustrating prehistoric birds. Almost all of the characteristics we use to distinguish fossil species from one another are skeletal, and most of those features would have been completely obscured in life by feathers and other soft tissue. Bird species can almost always be distinguished visibly by markings, crests, coloration, or behavior, all things which fossilize only very rarely.

Imagine that a time travelling paleontologist arrives in the Judithian age of Alberta 75 million years ago. She observes two similar but obviously different species of long-tailed ground birds in the wild, but they superficially differ only in their coloration. She knows that six long-tailed ground bird species are known to have lived at this time and in this place: caenagnathiformes *Chirostenotes pergracilis* and an unnamed species of avimimid, eudromaeosaurians *Dromaeosaurus albertensis*, *Hesperonychus elizabethae*, and *Saurornitholestes explanatus*, and troodontids *Troodon formosus*. Some of these could be easily ruled out: the caenagnathiformes would be obvious with their small, beaked heads, stout tails, and lack of retracted second toe claws. *D. albertensis* could probably be distinguished by their tall, square snouts. *H. elizabethae* would likely be differentiated by their small size and large wings, and likely by the presence of hind wings. *S. explenatus* and *T. formosus*, however, would be close in size (especially if the *T. formosus* were immature). The feet and lower legs of *S. explanatus* would be broader and more robust, and the snout of *T. formosus* would be more pointed at the tip as seen from above, but these details would be difficult to spot without close examination. The troodontids may or may not have more extensive, broader tail fronds compared to the eudromaeosaurians, which tend to have fronds expanded at the tail tip, but this kind of inference couldn't be used to reliably distinguish species without prior confirmation. To confidently identify these species, the paleontologist would need to capture and examine or even dissect several specimens to match skeletal anatomy with newly found and externally visible differences in coloration or feathering.

This scenario shows some of the inherent problems with creating a field guide to extinct animals. However, while informed by an abundance of speculation, such guides can be useful in illustrating our general view of the known diversity of extinct animals while pointing out plausibly identifiable distinguishing features. In some cases, I have deliberately over-generalized known traits from a few species to entire groups. For example, some troodontids (*Anchiornis huxleyi, Jinfengopteryx elegans*) preserve fronds of tail feathers which cover nearly the entire skeletal tail down to the base. While we can't be sure this pattern held true for all troodontids, and while it's extremely likely that exceptions existed, I have restored most troodontids with extensive tail fronds. Similarly, most other known ornithodesmid and basal avialan fossils have short fronds restricted to the tail tip; again, I carried this over to all eudromaeosaurians as a unifying trait to aid identification.

Some features, like the distribution and types of feathers, beaks, teeth, and even coloration, can be more confidently inferred based on evolutionary relationships and ecology: this process is explained in the sections below.

Above: Troodon formosus *(top) and* Saurornitholestes explanatus *(bottom) restored to the same length with identical feathering and coloration.*

Feathers & Wings

Feathers are classified based on their anatomy, their location on the body, and their function. There are two primary types of feathers. Pennaceous feathers have a rigid central filament or "quill" called a rachis (plural: rachides), from which stems numerous smaller filaments called barbs, linked together into a coherent vane by tiny barbules and microscopic hooklets. This arrangment of barbs, barbules and hooklets allows the vane to separate and to be "zipped" back together in a manner similar to Velcro. Plumulaceous or "down" feathers have short central filaments and barbs that lacks barbules and hooklets, preventing the formation of vanes and rsulting in soft tufts. Down feathers are usually present as an insulating layer below the contour feathers of a bird. Pennaceous feathers of the wing are called remiges (singular: remix), and those of the tail are called rectrices (singular: rectrix). Body feathers are called contour feathers, and are generally more loosely arranged and softer, though still vaned and pennaceous.

In Mesozoic birds as in modern birds in general, the feathers on the bottom of the head often began at or before the same point as the upper feathers, and were often long, creating a round-headed appearance. Similarly, most feathered fossil birds show very long feathers on the neck. Though long depicted as sinewy and slender, the necks of most maniraptorans were buried in feathers and would have appeared very short when not fully extended.

In most modern birds, the metatarsus and toes are bare and scaly, with scutes covering the top of the foot. This seems to have been the case in most Mesozoic birds as well, except for the most primitive. Many of the basal-most birds had long feathers on their legs and shorter feathers covering the toes. It may be that the first feathered animals were completely covered in feathers and subsequently re-developed scales only on the legs. Alternately, feathered legs may have been a genetic consequence of the development of hind wings in some species.

It is common to refer to bird-like dinosaurs or early birds having "arms" and "hands". However, many of these species possessed fully formed wings. Rather than being something present "on" the arm or hand, wings incorporate the entire forelimb. The primary feathers of the

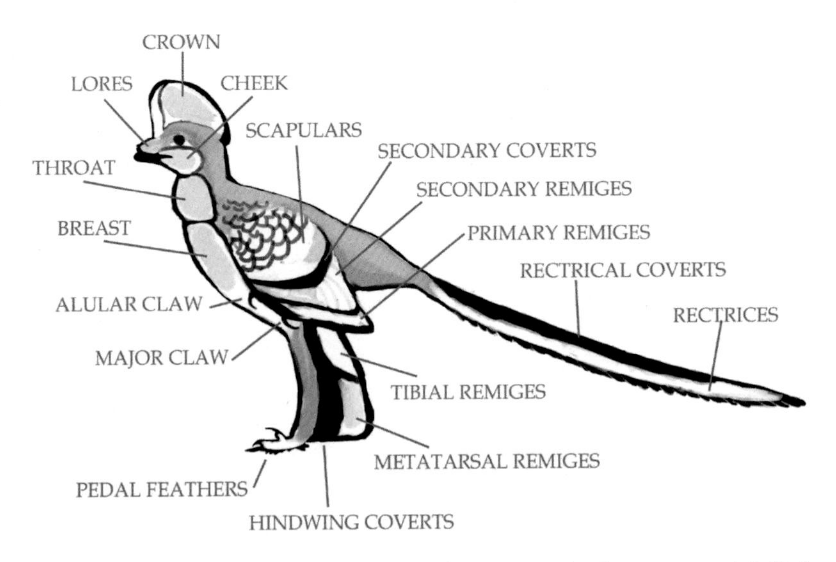

Above: Illustration of Anchiornis huxleyi, *with major feather anatomy labelled.*

wing anchor to the the first few phalanges and the metacarpals of the forelimbs, and so are as much a part of the "hand" as the claws are. Ligaments anchor the wing feathers to the muscles and bones of the wing, often leaving traces in the form of quill knobs, or ulnar papillae. These ligaments allow the feathers to be moved or folded in relation to the bones of the wing. This motion is usually accompanied by folding the wrist backward, a motion made possible by the presence of a half-moon shaped wrist bone called the semilunate carpal. The degree to which the wrist could fold varied among primitive birds, only becoming truly tight-folding in powered fliers. While the exact folding angle is difficult to calculate for fossil species, some researchers have been able to provide maximum estimates based on factors such as the angle of the radiale bone in the wrist (Sullivan et al. 2010). The most primitive paravians (including archaeopterygids, basal troodontids, basal ornithodesmids, etc.) could fold the wrist to about a 90-100 degree angle relative to the ulna (Senter 2006). This angle was reduced in some flightless lineages like the eudromaeosaurs. *Deinonychus*, for example, could only fold their wings at about a 120 degree angle, a modification possibly due to greater use of the wing in predation (Senter 2006). The folding angle increased dramatically in flighted lineages leading to modern birds, which often achieve a 60 degree angle between the metacarpus and ulna. Interestingly, at least basal caenagnathiformes were able to fold their wrists to a degree even more extreme than most modern birds (Sullivan et al.

2010).

Birds in general have three fingers, or digits, in the hand (manus). The first, shortest finger is called the alular digit (so named because it anchors the alula in modern birds, discussed below). The second, largest and longest digit is called the major digit. The third finger is called the minor digit. The minor digit is usually long and slender in primitive birds, but in many powered fliers, it became reduced and, in modern birds as well as some extinct lineages, fused to the major digit. It is likely that the minor digit was joined to the major digit by skin and soft tissue even in some more primitive birds. Claws (unguals) are a primitive trait for birds--the ancestral bird had claws on each of its three digits. The relative sizes and curvatures of the wing claws varied considerably among early birds, and they may have been used for everything from preening to climbing to predation and fighting. In more advanced birds, some or all of the wing claws were lost. This reduction of wing claws occurred in at least two separate lineages (enantiornitheans and euornitheans), both times in species with increased flapping ability. It's possible that most primitive avialans used their wing claws primarily for climbing up sufficiently high to launch into the air. When flapping flight advanced to the point that it was possible for these birds to launch from the ground, the role of the claws was reduced. Claws may have been retained as vestigial organs in some lineages; in fact, even some modern birds retain small claw sheaths on their alular digits.

The outer wing is composed of primary feathers, which attach to the major digit (both the metacarpals and phalanges). In flying birds, the vanes of these feathers are typically the most highly asymmetrical, differing in breadth on either side of the quill. In many birds, the primaries overlap each other to create a smooth border; however, in some species the outer primaries are separated or notched for greater ma-

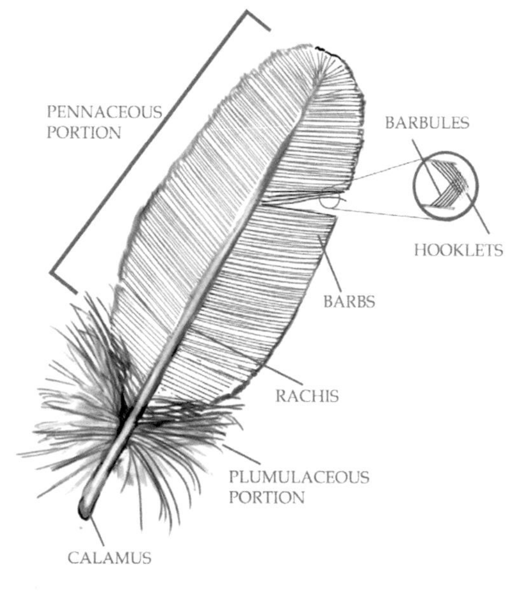

Above: Basic feather anatomy.

neuverability. In some primitive birds (those without a fused hand, or carpometacarpus), the flexibility of the digits in the wing could have allowed for additional maneuverability by altering the wing's curvature and the spacing of the feathers.

Secondary feathers, which attach to the rear forearm bone (ulna), are usually broader and more rounded than primaries. The secondaries typically form the entirety of the inboard wing, and point slightly toward the body as they approach the elbow. This gives the inner wing a rounded profile and helps bridge any gap that forms between the body and the wing due to short or missing tertials.

Tertiary feathers, or "tertials", are present in some birds, in which they attach to the upper arm (humerus). Most birds generally keep the upper arm mostly pressed against the body even when in use, and so their tertials are either absent or reduced to contour feathers similar to those found on the body. Only specialized soaring birds, which extend the entire arm into a nearly straight line or maximum length, have tertial feathers which are similar in appearance to the secondaries. Only some avians are known with certainty to have had tertiary remiges, though

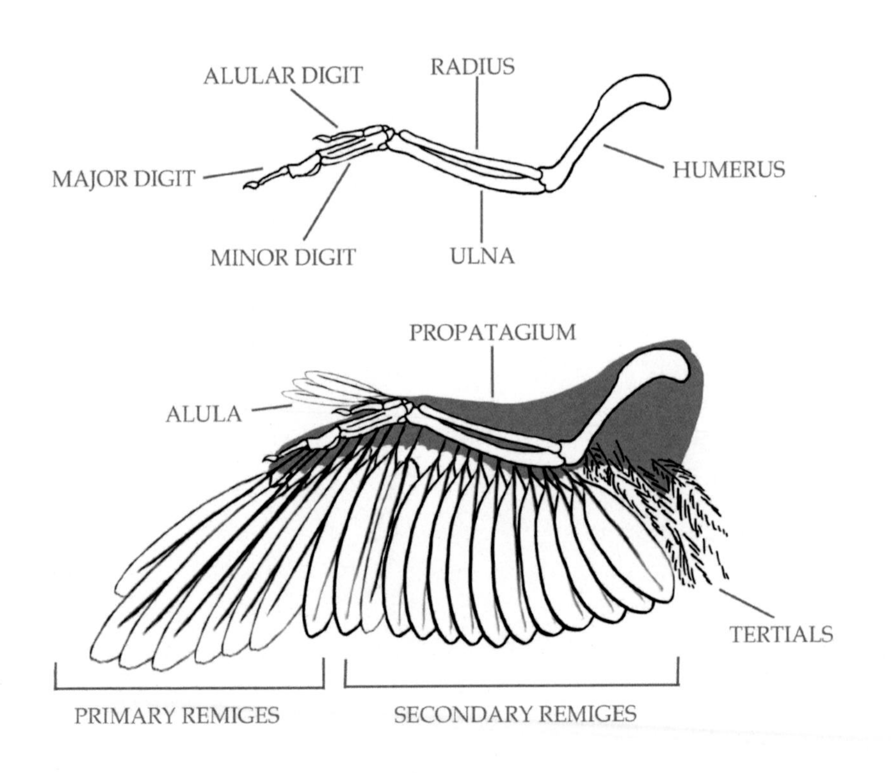

Above: Wing anatomy of a generalized non-avian avialan bird.

they may have been present in other euornitheans, such as Ichthyornis, which were probably gull-like dynamic soaring birds. More primitive birds like *Archaeopteryx* lacked tertiary remiges but had contour-like feathers (as well as scapulars) partially filling the gap between the wing and the body.

The gaps between the arm and the primary, secondary, and tertial wing feathers (together called remiges) were covered with several layers of smaller feathers known as coverts. In all but the most primitive birds, the minor digit also anchors pennaceous feathers, in a structure called the alula which aids in aerial maneuverability and breaking.

Additional covert-like feathers attached to the shoulder, called scapulars, help fill the gap between the secondaries and the body and partially cover the remiges when the wing was folded away. The scapulars are responsible for the effect that most of the wing blends smoothly into the body feathers in life when a bird is at rest, typically with only the remix tips protruding. This effect would be lessened in primitive birds which had wings incapable of folding tightly against the body, and the remiges would have been more conspicuous at rest. In advanced euornitheans, the wings are situated high on the body, so that the upper arms (humeri) sit parallel to each other on top of the back. This causes the folded wing feathers to form a 'cloak' across the top of the bird's body. In more primitive birds, many of which had wings situated low on the body with shoulders near the breast, the wings would have folded against the sides of the body, leaving the contour feathers of the back exposed.

In all known birds, including those primitive forms such as *Microraptor*, the wrist was connected directly to the shoulder via a stretch of skin and ligaments known as the propatagium. Covered in feathers and blended into the body of the wing, this structure prevented the elbow from extending in a straight line, and smoothed the front of the wing despite the fact that the elbow was habitually held at a V-shaped angle (which also forced the upper arm to be held more or less against the body).

The overall shape of the wing is determined by the length of the wing bones combined with the lengths of the remiges. Most primitive birds had broad, rounded wings of a low aspect ratio, with relatively short primaries. Exceptions to this rule include *Microraptor*, *Confuciousornis*, and *Hongshanornis*, which had very long primary feathers giving them high aspect ratio wings. Generally, short, broad wings are equated with higher maneuverability and are often found in forest birds which fly in short bursts and sharp turns between trees and among shrubs.

Long, pointed wings are often found in soaring birds such as gulls, or those that fly at high speeds in the open like swifts. Tellingly, most enantiornitheans, which were almost exclusively arboreal forest birds, had short-round wings regardless of how advanced their anatomical flight apparatus had become. Most primitive euornitheans were ground birds or generalized water birds and, like modern fowl, retained relatively broad, rounded wings, except for some specialized types, like Hongshanornis, which flew mainly over open lakes and other waterways. In this book, those birds for which the wing shape is unknown, like *Ichthyornis* (marine birds restored with long, pointed wings of a high aspect ratio) are given wing shapes inferred based on their ecology and phylogenetic relationships.

Beaks & Teeth

In modern birds, the snout is covered in a keratinous beak (known scientifically as a rhamphotheca), and therefore is bare of feathers. However, many Mesozoic birds lacked beaks and thus the extent of the snout feathering was highly variable. Few fossils preserve the delicate and usually short feathers present on the snout, however, we know of some species, such as *Eoenantiornis buhleri* and *Sinornithosaurus millennii*, in which the feathers extend three-quarters of the way or more toward the tip of the jaws, resulting in a snub-billed appearance. In other species, such as *Microraptor zhaoianus* and some enantiornitheans ("opposite birds") with specialized and elongated jaws, the feathers did not extend much forward of the eye. As in modern birds, it is likely that some had bare heads and/or necks. Featherless portions of the head in modern birds are usually related to display, and in some examples, heat loss or ease of preening.

It is well known that many dinosaurs had beaks, but also that, famously, many of these also possessed teeth. However, the exact arrangement of beak and teeth in the jaw is commonly misunderstood. Almost any life restoration of a *Hesperornis*, for example, will show a keratinous beak covering the entire extent of the upper and lower jaws. Some of these clearly show teeth erupting directly from the tomia (edge) of a continuous keratin beak. The continuous appearance of this beak is likely incorrect in itself, since non-avian birds probably all had "compound rhamphotheca," beaks made up of several distinct plates that are often visible in life.

In bird species like *Hesperornis regalis*, the lower jaw (dentary) teeth continue almost all the way to the jaw tip, though the very tip (and the small predentary bone that was probably present) was toothless. On the underside of the upper jaw (premaxilla), there were indentations where the lower teeth would have locked into the bone. If there was a hard beak present, it would have had to have been pitted to accommodate the lower teeth. However, these indentations are inset to the edge of the jaw. The edges of the upper jaw slightly overhang the lower jaw, which would have allowed for the beak edge, if it was there, to not come into contact with the lower teeth, which would have caused tooth wear any time the mouth closed. Upper teeth in this species are restricted to the very back of the mouth (maxilla). This can also be seen in the 'dental grooves' on the underside of the skull.

Above: Evolution of the avian beak and loss of teeth. From top to bottom: Archaeopteryx lithgraphica, Yixianornis grabaui, Hesperornis regalis, Polarornis gregorii. *Not to scale.*

According to Heironymus & Witmer 2010, in both *Ichthyornis* and *Hesperornis*, the premaxillary nail and mandibular (lower jaw) nail were the most heavily keratinized parts of the beak. These "nails," which often formed small hooks at the beak tips, are where the beaks would have been most solid, like typical modern bird bills. The same authors note that the simple presence of teeth in the maxilla and dentary of these species probably means that they entirely lacked the latericorn and ramicorn beak plates which normally cover the 'lips' of the jaws, and that the presence of hardened rhamphotheca on the edges of the jaws may be unique to modern birds. However, as noted above, the tip of the upper jaw (premaxilla) in *Hesperornis* is also toothless and provides space for an overhanging edge (tomia) of some kind to be present. This would have been somewhat softer tissue, like the more pliable bills of ducks and geese. Further support for the presence of a beak on the premaxilla comes from the presence of a "rhamphothecal groove" on the upper part in front of the naris (nasal opening in the skull), which likely served as an anchor point for the keratin on the skull.

So how far did the beak extend? Heironymus & Witmer found that the latericorn almost always extends to the back of the subnarial bar in birds. This is a process of the premaxilla that extends back to separate the naris from the maxilla. This means that beaks will very rarely, if ever, extend onto the maxilla itself. The maxilla in *Hesperornis* even compensates for this limitation by extending a bit forward underneath the subnarial bar to extend the tooth row past the full extent of the beak.

Based on the evidence above, in *Hesperornis,* the toothless, pointed beak tips would have been made of solid, normal keratin, while the rest of the beak would have been more like stiffened skin grading into normal skin and feathers toward the back of the skull. At no point

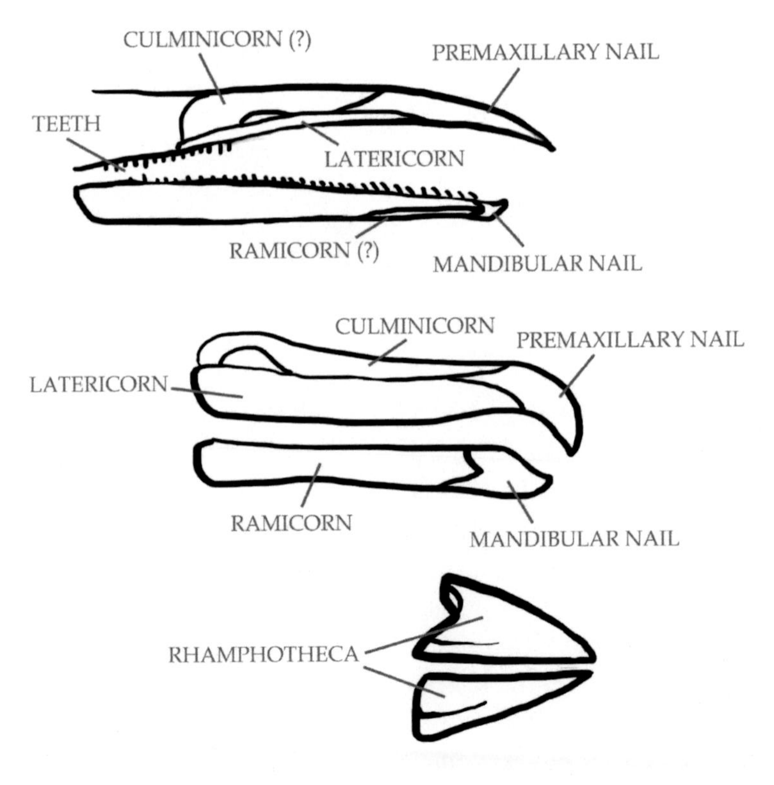

Above: Anatomical terms for the parts of compound beaks. From top to bottom: A hesperorn, an albatross, and a finch. Not to scale.

would the teeth have occupied the same physical space as the rhamphotheca, though they may have overlapped thanks to the fact that parts of the tooth row were inset to the jaw. The rhamphotheca never seems to have housed tooth sockets, so the beak and the teeth were effectively segregated in different parts of the jaws. *Hesperornis* is perhaps the most well-studied case of a bird with both a beak and teeth, but there is no reason to suspect that the same general principles would not have held true in other beaked and toothed birds. In short, no Mesozoic birds had "teeth in their beaks" as is often stated and depicted in art. Rather, they had both beaks and teeth in different parts of the skull, presumably serving different roles in food capture and processing. A tooth protruding from the beak, relegating the keratin itself to essentially the gums, would have rendered the beak useless anyway. Teeth protruding from a beak would have been a redundancy, an expense that would not have been evolutionarily advantageous.

By using phylogenetic bracketing, it is possible to construct a rough picture of the distribution of beaks among Mesozoic birds. Beaks do not seem to have been present in any of the first bird lineages, likely because both early avialans (gliding birds) and early deinonychosaurians (sickle-clawed birds), while some seem to have dabbled in omnivory, were mainly carnivorous. However, many deinonychosaurian fossils which preserve feathers show a small portion of the tip of the snout that is unfeathered. This featherless snout tip is also seen in some toothed, beakless enantiornitheans. It is possible this could be evidence of "rhamphotheca" in its loosest sense--the very lightly cornified, flexible bill skin found toward the back of the beaks in some modern birds, where the horn-like, keratinous portion thins out into normal skin.

Around the base of *Avialae*, the reduction of teeth becomes commonplace in several independent lineages, probably due to a shift to more omnivorous diets. Almost all known basal avialans have very few, very small teeth in the upper jaws, and several lost teeth altogether. This trend appears to culminate with the confuciusornithids, which are not only toothless but have sharply pointed jaws that, in some very rare specimens, preserve the actual keratin of a beak. These impressions show that in early beaked birds, the rhamphotheca was thin and delicate and probably not as heavily keratinized as in modern birds.

In most enantiornitheans, the jaws are fully toothed, with no evidence of beaks. It may be tempting to think that this could unite the enantiornitheans with the toothy, beakless deinonychosaurians in a "Sauriurae" to the exclusion of the beaked euornitheans ("true birds").

However, given the numerous times beaks have evolved independently in vertebrates, it's more likely that each of the examples of basal avialans with reduced or absent teeth arose independently of one another, or that some reversal occurred at the base of ornithothoraces to return birds to a state of fully-toothed jaws. While many enantiornitheans preserve jaw material, only one species exhibits the kind of toothlessness at the front of the jaws that could imply a beak: *Gobipteryx minuta*, which, like confuciusornithids, were beaked and completely toothless.

All known Mesozoic euornitheans (the fan-tailed birds, including modern birds), unlike the typically beakless enantiornitheans, had small beaks restricted to the jaw tips, with teeth further back in the jaw. While hongshanornithids were originally reported to have beaks and to be completely toothless, O'Connor and colleagues later showed that they had tooth sockets preserved in the upper and possibly lower jaws. However, the jaw tips were toothless and probably beaked. In some (perhaps most) Mesozoic euornitheans, an additional bone was present forward of the dentary: this predentary bone was always toothless and likely evolved specifically to accommodate a jaw-tip beak. Predentaries are known from the most primitive euornitheans like *Hongshanornis longicresta* up to the most advanced non-avian species like *Hesperornis regalis*; however, they seem to have been lost or incorporated into a solid, single lower jaw bone shortly before the advent of modern birds. Interestingly, the only other group of vertebrates to have evolved predentaries are the ornithischian dinosaurs, which had a similar arrangement of beaked and toothless jaw tips in front of fully toothed jaws.

The earliest fully beaked and toothless euornitheans are also among the most primitive: *Archaeorhynchus*, which lacked teeth and had flattened, spoonbill-like beaks. Because more advanced birds retained teeth in both jaws, this is almost certainly an independently-acquired condition unrelated to the toothlessness of modern birds. The songlingornithids and the later hesperornitheans and *Ichthyornis* all had toothless premaxillae and predentary bones with toothy maxilla and dentaries. Evidence from bone texture shows that they likely had keratinous beaks at the tips of their jaws, and either feathery toothed jaws or pliable, skin-like rhamphothecae posterior to the beak. Since both major lineages of modern birds (avians) lack teeth, it's probable that their common ancestor was also fully beaked, so teeth must have been lost for good in the bird lineage shortly after ichthyornitheans evolved. Interestingly, studies of ichthyornithean and hesperornithean bone structure shows that they likely had "compound rhamphotheca", and this may have been the ancestral condition for modern birds (Heironymous & Witmer, 2010). While the quintessential bird beak is made up of a single keratinous sheet covering the jaw, in species with compound beaks, the keratin is arranged in discrete plates on the jaws. This can best be seen in some modern seabirds like the Albatross.

Unfortunately, in interesting groups like *Hollanda*, *Gansus*, and *Patagopteryx*, the condition of the jaws is unknown. However, we can use parsimony and phylogenetic bracketing to try and develop an educated guess. Most studies find these three groups to be euornitheans ("true birds") more primitive than the hesperornitheans and *Ichthyornis*, which have toothed upper and lower jaws with beaked tips. While the three intermediate birds may well have lost some or all of their teeth independently again, all things being equal it is more parsimonious to suppose that like ichthyornithids, hesperornithids, hongshanornithids and chaoyangornithids, they had beaks restricted to the tips of the jaws, with teeth behind.

Feather Color

Recent work by Jakob Vinther and others on reconstructing the life coloration of prehistoric birds has been some of the most exciting paleontological research of the decade. Prior to this research, artists were often considered to have had total license to imagine the external appearance of prehistoric dinosaurs. However, even without direct measurement of colors and color patterns in fossil species, there are certain biological factors which go into bird coloration which have been largely ignored by artists in the past.

There are several processes that add color to feathers. At the most basic level, these can be categorized as either structural color or pigmentation, though often these two modes combine to create the life coloration of a bird.

Structural colors come from the actual physical structure of the keratin or melanin in the feather. At the microscopic level, many feathers exhibiting structural color have a "foamy" texture of tiny spheres or channels which enclose minute air bubbles. Light scatters through these bubbles in various ways depending on their exact arrangement. The development of these complex structures has recently been examined by

Dufresne & al. (2009). Alternately, the structure may be produced by the layering or physical arrangement of melanin granules in the feather (Stettenheim, 2000).

Structural colors can have two effects on the life appearance of an animal; they can produce colors not found among the various pigments, and enhance or change pigment colors. For example, among amniotes (vertebrates which lay shelled eggs), there is no known method of blue pigmentation. Blue skin, scales and feathers are produced by light scattering due to structural configuration. Similarly, iridescence as seen in many birds comes from the feather structure. A bird with bright white or pitch black feathers likely uses structural colors in addition to pigments (or lack thereof) to achieve this effect--without them, these colors would be flatter, duller, and less vivid. Structural coloration can also act as a filter, modifying the light reflected by pigments to form new colors. In most birds that have them, green feathers are produced by layering yellow pigmentation nodules over a blue-producing underlying structure.

Though harder to find in the fossil record than pigments or chemical traces, structural color can be found in some fossil feathers. Iridescent feather fossils have been reported by Vinther & al. (2008), and it is sometimes apparent even to the naked eye. Structurally colored feathers have been recognized by a distinct arrangement where a thin layer of densely aligned melanin granules overlies a looser conglomerate of melanin. This can be seen even if the overlying keratin scattering layer has degraded (Vinther & al. 2008). This arrangement where structure is produced by the arrangement of melanin as well as 'bubbles' in the keratin layer is notably found in the dazzling iridescent plumage of hummingbirds (Prum, 2006).

The mechanics of structural color in feathers have implications for how extinct bird species are restored in art. Blue, green, jet black and bright white can't be present in birds that lack structural color in their feathers. Structural colors may or may not have been possible in the monofilament feathers of some primitive coelurosaurs and ornithischians. Note that structural coloration is never observed in the monofilament hair of modern mammals. The primary difference between hair and simple feathers, however, isn't the macrostructure of the filaments, but the microstructure of the underlying molecules. Hair is composed of alpha-keratin, a helix-shaped molecule like DNA. Beta-keratin, which makes up feathers, has a layered and pleated underlying molecular structure more conducive to scattering light. On the other hand, in all of the iridescent fossil feathers studied by Vinther & al. (2008), the structural color was restricted to the barbules, which are not present in many primitive feathered dinosaurs. Additionally, structural colors are not observed in modern plumulaceous feathers (down) or in the downy after-feathers of otherwise structurally colored pennaceous feathers. It is therefore likely that blue, green, iridescent or vivid downy and monofilament feathers were extremely rare, if they existed at all in Mesozoic birds and more primitive feathered dinosaurs.

The vast majority of bird colors are due in whole or in part to pigmentation, or lack thereof (Stettenheim, 2000). There are several different kinds of pigments, with the two most common being melanins and carotenoids.

Above: Hypothetical restoration of a climbing subadult ornithodesmid Deinonychus antirrhopus. *Non-neoavian bird coloration was probably limited to earth tones and iridescence due to the inability to sequester carotenoids in the plumage.*

Melanins are easily identified in fossil feathers, and their shape and concentration can indicate what color they produced. Melanins are responsible for black (though not deep, solid black, which requires the addition of structural color), gray, and a wide variety of browns to rufous orange or rusty red colors. A lack of melanin will produce white, as evidenced by albino specimens. Note that some albino birds today are not completely white, but retain some darker coloration due to the structural colors of the feathers which are not erased by an absence of melanin in all cases.

Carotenoids are, by and large, what give birds their characteristically bright colors. Carotenoids cannot be directly synthesized by the body in most animals (some can do this, but there need to be other types of carotenoids present to convert). Carotenoids come almost exclusively from a diet of plants or, secondarily, of things that sequester a lot of carotenoids in their body tissues (like plant-eating invertebrates and some fish). Gulls living near salmon farms have shown hints of pink in their feathers: this is because farm-raised salmon are fed artificial carotenoid sources to make their flesh pink, and these are transferred to the birds. The most unusual source of carotenoids, this time among a carnivorous species, is the Egyptian Vulture, which gets its bright yellow facial skin by eating the dung of ungulates, which yield no significant nutritional value and appears to be consumed by the vultures only for its carotenoid content (McGraw 2006). Indeed, while carnivores aren't usually brightly colored, there may be selective pressures in some species to add unusual supplements to the diet in order to become more colorful (McGraw 2006).

Carotenoids are often used by modern birds as a sign of fitness when choosing a mate. Because carotenoids have to be eaten, a bird with a poor diet will be drabber than a bird that is very successful at finding food. A flamingo kept in a zoo will turn white if its diet isn't artificially supplemented with red carotenoids. Carotenoids can also impact the eye color of a bird, as well as beak color and the color of the scales on its feet: even the yellow yolk of a chicken egg (Zongker 2007) is due to carotenoids (some birds use Flavin for yolk color, which will be discussed later in this chapter).

Note that even modern birds do not have fine control of carotenoid pigmentation in their feathers. Carotenoids are almost always found coloring large swaths of feathers, not as small spots or intricate detailing within individual feathers (Hill 2010).

Carotenoids have so far not been reported in fossils, primarily because carotenoid granules look the same as melanin granules (melanosomes), and unlike melanin, carotenoids cannot be distinguished by shape. According to Li & al. (2009), special chemical tests could be run to determine if a melanosome is really a carotenoid, and what color it was. The chemical-based analysis of feather color patterns conducted by Wogelius & al. (2011) brings us another step closer to being able to identify cartenoid-based coloration in fossil birds.

Even if we could currently test for the presence of carotenoids, it is uncertain whether or not most Mesozoic birds would have been able to use them as feather pigment the way modern birds do. The biological ability to sequester carotenoids in the feathers appears to be absent from birds of the subgroups *Palaeognathae* (the ostriches, emu, kiwi, tinamou, etc.) and *Galloanserae* (including ducks, geese, pheasants, etc.). While these birds can and do use carotenoids to color the skin of the feet, face, or

Above: Illustration of Jeholornis prima.

bill, they seem to lack the chemical pathways necessary to transfer carotenoids into the feathers (Hill 2010). Unless this represents two successive evolutionary reversals, it is probable that colorful, carotenoid-pigmented feathers are unique to the modern bird group *Neoaves*, in which case almost all Mesozoic birds could not have had bright yellow, orange, red, or green feathers, but would have been limited to iridescence and striking contrast to create effective visual displays. Artists should keep in mind that adding orange, yellow or green feathers, or red, orange or yellow beaks or skin, implies that a bird is eating a diet containing carotenoids, and that even this may not be reasonable for the feathers of non-avian birds.

While more rare than melanin and carotenoid pigmentation, porphyrin pigments provide another method of producing color in birds. Porphyrins are perhaps most famous for lending blood its red color and leaves their green (both heme and chlorophyl are varieties of porphyrin), but it can also color feathers, adding browns and reds as well as green, though green is only found in the specialized turacoverdin variety found in Turacos. Interestingly, porphyrins may play a role in temperature regulation. In addition to insulating eggs (see below), they are mainly found in the downy feathers of nocturnal birds like owls, and those that are active at cold temperatures. Another reason porphyrins are found mainly in non-pennaceous feathers is that the compound makes feathers more labile, and so it would be detrimental to employ it in feathers which must stand up to mechanical strain and to the elements. This is opposite to the effect of melanins, which add strength and are often found at the tips of wing feathers where stresses are high.

Porphyrins are often responsible for producing the blue of American Robin eggs, and most other egg coloration. In fact, some researchers note a correlation between porphyrin in eggshells and nesting behavior. Pure white eggs are only found in birds which nest in shelter, such as under dense foliage, and which constantly attend to their eggs. Species which leave their eggs partly exposed to the elements have colorful porphyrin-containing shells, partially for reasons of camouflage, but also possibly due to the supposed temperature-regulating effect of porphyrins.

Theoretically, it may be possible to detect porphyrin via chemical analysis of fossil birds. However, when restoring birds with no preserved feathers, there should not be much for artists to consider that is not already covered by a working knowledge of melanin coloration. Porphyrins in feathers produce mainly only brown and dull red, colors that could also be produced with melanin alone. If anything, porphyrins give artists license to add extra reddish splashes to purely carnivorous species, especially those that may have been active at night or in cold climates.

There are a variety of minor and less common pigments that can color a bird's feathers. Pterins are responsible for the yellow, red, white, and orange colors of some bird eyes (in humans, eye color is controlled by melanin; low melanin results in blue eyes, and some babies' eyes darken as their melanin levels increase). Flavin pigments cause many egg yolks to be yellow. Psittacofulvins are found only in some parrots, and create yellows, oranges, and reds in place of carotenoids, which parrots have evolved to sequester, possibly for nutritional reasons. There are (currently undescribed) pigments known only in penguins that add fluorescence to their yellow display feathers.

In conclusion, Mesozoic birds would have taken their coloration predominantly from melanin and iridescent structural coloration. This would have resulted in a bird fauna possibly more drab overall than the one we see around us today, due to the comparative lack of neoavians and their ability to color their feathers with carotenoids. However, like modern ducks, these birds would still have been able to create intricate and dazzling displays of color and pattern using combinations of earth-toned melanins (muted yellow, rusty red, dark grey and off-white) as well as layering of iridescence to create striking, jewel-like feathers in blue, green, purple, and glossy black and white. While they wouldn't have the pinks of flamingos or bright greens and yellows of birds-of-paradise, Mesozoic birds may have been just as beautiful.

Guide to Mesozoic Birds

About the Guide

Each entry in this guide consists of a brief caption outlining the species paired with one or more illustrations of how specimens may have appeared in life. The reconstructions are necessarily speculative to varying degrees (informed by the principles outlined above). Very fragmentary species are illustrated only if educated inference for their life appearance can be drawn, e.g. from phylogenetic bracketing (filling in the gaps based on known close relatives, or inferring likely appearance based on their classification). Species which are very fragmentary and for which no good inferences about appearance can be made are not illustrated, but are listed in Appendix A. Identification marks point to known areas of anatomy which would be used to distinguish species in life, such as features of the snout, legs, wings, or feathers.

Each illustration caption contains the following information:

Etymology/Common name:
Fossil species usually do not have common names, so these are drawn from the etymology of the scientific name. Translations defer to the original etymology as given by the naming author where appropriate (e.g. -saurus, which has been variously translated as "reptile" or, more correctly, "lizard").

Scientific name:
The generic and specific names of the bird. Species which have not yet been formally named are noted as such.

Time:
Approximate age of the species, in Mega anni (Ma, or millions of years) before present.

Location:
State, province, or region, and country of origin.

Habitat:
The specific geological formation(s) from which fossils have been recovered, with brief descriptions of what the the ecosystem(s) may have been like in life based on fossil and other geologic evidence.

Size:
Estimated wingspan (WS), body length (skeletal; BL) and total length (i.e. including rectrices; TL) are given where possible in metric and American units. If relatively complete wings are known but lack complete primary feathers, wing span is given as >(arm span).

Features:
Description of any distinguishing features which would have been visible externally based on known fossil evidence or inferred based on related species. Some species are distinguished only based on anatomy which would not have been visible in life without dissection. These are reconstructed with speculative variations in color, plumage, or other soft tissue features.

Biology:
Description of any biological or behavioral features that can be reasonably inferred, including differences between various growth stages, differences between sexes, behavior/interaction with the environment, and diet.

Species are arranged into major groups, which are organized in roughly chronological sequence, though minor groups ("subfamilies" etc.) are placed together where possible for purposes of comparison. Poses are standardized in either slow-walking or standing/alert postures. For species inferred to be highly arboreal, a perching pose is used. Dorsal views are provided to illustrate wing shape for those species that may have been able to fly or glide. For species that exhibit modifications or reductions of the teeth, presence of toothlessness and/or a beak, or other notable features not visible in a lateral standing posture, additional anatomical illustrations are provided.

Note that the individuals illustrated together on each page are not drawn to exact scale, though where possible they are drawn to approximate relative sizes. Scale diagrams are provided with each species for comparison.

Basal Caenagnathiformes

The first highly diverse offshoot of the early Mesozoic bird lineage are a group of bizarre, omnivorous ground birds called the carnagnathiformes ("recent jaws", alternately oviraptorosaurs). The most primitive types, like *Caudipteryx*, were long-legged and fairly small-winged, with only a small number of teeth, or beaks. More advanced carnagnathiformes became larger, and some had elaborate casques similar to modern hornbills or cassowaries. At least one grew to enormous sizes: *Gigantoraptor erlianensis*, at up to 1.4 tons, are the largest birds of all time.

Most researchers, based on cladistic analysis, find the caenagnathiformes to be more distantly related to modern birds than is *Archaeopteryx lithographica*. Despite this, they share some strikingly bird-like features that must otherwise be explained by convergent evolution, including nearly toothless jaws and shortened tails with fused vertebrae at the tips.

Famously, several specimens have been found brooding their nests as modern birds do, indicating that some bird behaviors likely evolved before or concurrently with the advent of true feathers. In part due to the characteristics of the most primitive known species *Protarchaeopteryx robusta*, many researchers had speculated that carnagnathiformes were close relatives of the segnosaurs (also known as therizinosaurs), bizarre bird-like herbivorous dinosaurs with distinctively huge, scythe-shaped claws on their hands. However, contrary to what would be expected, impressions of feathers from segnosaurs showed only down feathers and simpler quill-like filaments, unlike the true vaned feathers of caenagnathiformes. This, as well as some more detailed phylogenetic analyses, has shown segnosaurs to be more primitive than true, wing-bearing birds.

The caenagnathiform diet has remained largely mysterious. Only the most primitive species had teeth, and most later groups were beaked. Some fossils preserve gastroliths in the stomach, suggesting at least partial herbivory, while others have preserved the remains of small lizards in the stomach contents. The beaks of most species were stout and strong, resembling those of parrots or turtles.

The wings were in many cases able to fold more tightly against the body than those of avialans. Despite this, the wings were generally small and all known species were flightless and probably primarily terrestrial. The tails, while short, were extremely strong and flexible, and this unusual range of motion was probably employed in mating displays.

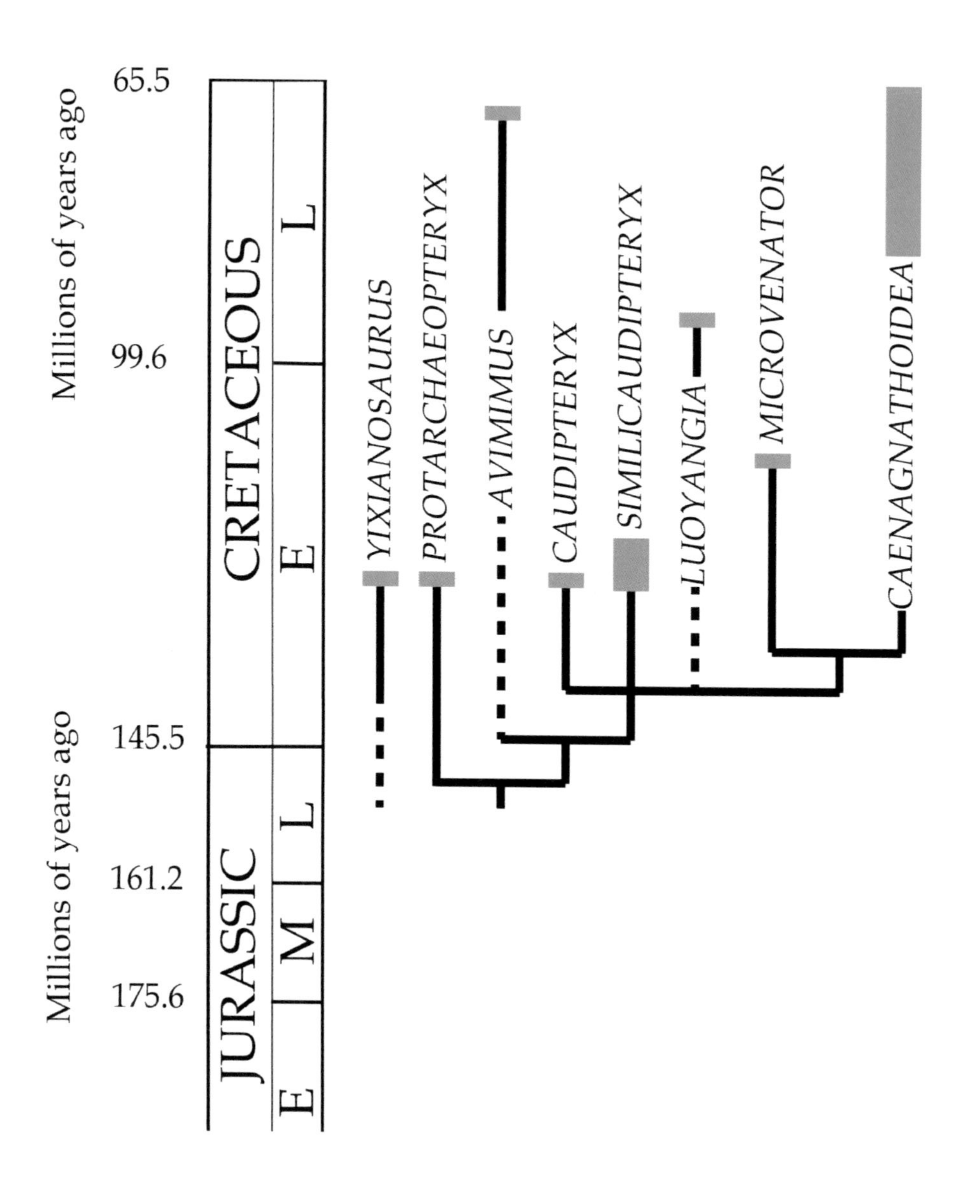

Above: Relationships of basal caenagnathiforms over time. Phylogeny approximated based on Senter 2007 and other sources.

Strong Primitive Archaeopter ***Protarchaeopteryx robusta***
Time: 124.5 Ma ago **Location:** Liaoning, China **Habitat:** Lower Yixian Formation (see above) **Size:** WS ~65cm (2ft); BL 55cm (1.8ft); TL 65cm (2ft) **Features:** Head short & high w/ rounded snout. Teeth numerous (up to 18 in both the upper & lower jaws). Upper front two teeth unusually large & chisel-shaped, giving a "buck toothed" appearance. Wings w/ large claws, but remiges unknown. Legs long. Tail frond small & square. **Biology:** The large, chisel shaped teeth indicate that these were probably specialized herbivores. The wings were relatively small, & the remix length unknown, but they may have at least allowed some limited parachuting ability, slowing descents on the rare occasions that they climbed or rested in trees (Chatterjee & Templin 2004). The contemporary species *Incisivosaurus gauthieri*, known from a complete skull, is probably a synonym.

Zou's Tail Feather ***Caudipteryx zoui***
Time: 124.5 Ma ago **Location:** Liaoning, China **Habitat:** Lower Yixian Formation (see above) **Size:** WS 70cm (2.2ft); BL 90cm (3ft); TL 1m (3.3ft) **Features:** Head triangular w/ narrow, pointed snout. Teeth restricted to tip of upper jaw; lower jaw toothless. Upper front two teeth large. Wings very small & lacking secondary remiges. Minor digit highly reduced, lacking a claw & probably fused to major digit via soft tissue. Alular claw slightly smaller than major claw. Legs long. Tail very short, w/ rectrical frond limited to the final third of the tail length. All rectrices nearly equal in length, creating a square- or diamond-shaped frond when seen from above. **Biology:** Presence of numerous small gizzard stones in some fossils indicates probable herbivorous diet. Tail frond light in color with a dense, dark banded pattern, suitable both for display and camouflage. Highly reduced wing feathers able to fold tightly against the body & may have also been used in display. Fossils of this species are common only in a relatively small region of the Yixian Formation, indicating very specific habitat preferences.

Dong's Tail Feather ***Caudipteryx dongi***
Time: 124.5 Ma ago **Location:** Liaoning, China **Habitat:** Lower Yixian Formation (see above) **Size:** WS 110cm (3.6ft); BL 75cm (2.5ft); TL 90cm (3ft) **Features:** Head long & relatively low & narrow, w/ teeth restricted to tip of upper jaw; lower jaw toothless. Upper front two teeth large. Neck long & slender w/ short feathers. Wings relatively large compared to other caudipterids. Secondary remiges long but restricted to the outer wing. Minor digit highly reduced, lacking a claw & probably fused to major digit via soft tissue. **Biology:** Similar to *C. zoui*, often considered synonymous. However, larger wings & presence of secondary remiges in a smaller specimen is opposite the growth pattern seen in *Similicaudipteryx*, so *C. dongi* may be a distinct species.

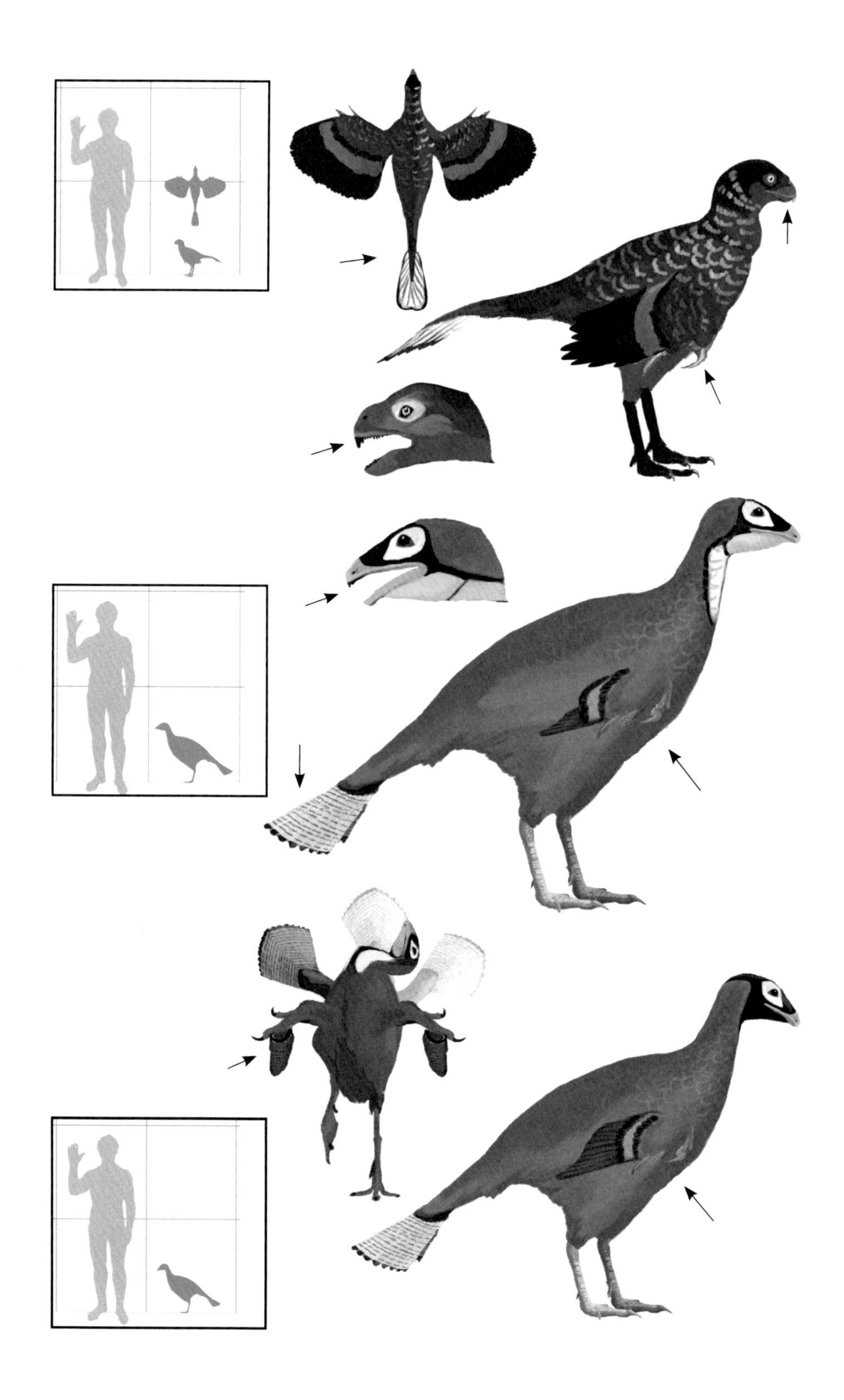

Heavy Tail Feather No scientific name (specimen number BPM 0001)
Time: 124.5 Ma ago **Location:** Liaoning, China **Habitat:** Lower Yixian Formation (see above) **Size:** WS 85cm (2.8ft); BL 75cm (2.5ft); TL unknown **Features:** Head long & rectangular w/ robust, blunt snout. Nasal opening relatively large compared to other caudipterids. Teeth restricted to tip of upper jaw; lower jaw toothless. Front two teeth large. Wings relatively small. Minor digit highly reduced, lacking a claw & probably fused to major digit via soft tissue. Alular claw slightly larger than major claw. Legs long. Tail very short. **Biology:** While details of the feathers are not well preserved in this species, it is clearly different from other caudipterids in the more rectangular shape of the skull, which is probably a primitive characteristic.

Yixian Similar Caudipter ***Similicaudipteryx yixianensis***
Time: 124.5 - 120 Ma ago **Location:** Liaoning, China **Habitat:** Yixian (see above) and Jiufotang Formation, cool, swampy marshlands. **Size:** WS ~ 90cm (3ft); BL 1m (3.3ft); TL ~1.3m (4.3ft) **Features:** Head high & rounded. Wings very large in subadults & adults. Wings in juveniles much smaller & lacking secondary remiges. Tail frond extremely large relative to body size & extending to the tail base in adults, unlike other known caenagnathiformes. Yixian specimens may represent a distinct species. **Biology:** Adult (Jiufotang) specimen possess fused, pygostyle-like tail vertebrae, indicating that the tail frond may have become even larger in mature individuals. The wings & tail frond were relatively small in juveniles, similar in proportion to *Caudipteryx zoui*. In immature specimens, the rectrices were longer than the primary remiges, & secondary remiges were absent. This indicates that the tail frond grew large first, & that the wings developed more slowly, likely being less important to juveniles. In adults, the wings & tail frond were more equal in size, but the rectrical frond was still larger than the wings, much larger than any other caudipterid. The significant size & late development of the wings & tail indicate that these structures were primarily used for display in this species.

Liudian Luoyang *Luoyanggia liudianensis*
Time: 96 Ma ago **Location:** Henan, China **Habitat:** Mangchuan Formation. Ecosystem dominated by freshwater lakes. **Size:** WS unknown; BL ~1.2m (4ft); TL unknown **Features:** Head long w/ broad, V-shaped beak & straight lower jaw. Internally, hip bone straight and flat rather than concave. **Biology:** Very primitive & similar to caudipterids, except for the toothless beak. Probably generalist omnivores.

Swift Small Hunter *Microvenator celer*
Time: 110 Ma ago **Location:** Montana, Oklahoma & Wyoming, USA **Habitat:** Cloverly Formation. Found in arid savannas dominated by ferns and low scrub and characterized by severe dry seasons. **Size:** WS >50cm (1.6ft); BL 85cm (2.8ft); TL unknown **Features:** Small toothless oviraptorids somewhat similar in appearance to the earlier caudipterids. Snout short & deep w/ square, toothless beak on lower jaw. Wings short. Legs relatively short w/ long femur & tibia but short tarsus. **Biology:** Possibly omnivorous, feeding on tough plants & small vertebrates in their arid environment.

Portentious Bird Mimic *Avimimus portentosus*
Time: 70 Ma ago **Location:** Omnogovi, Mongolia **Habitat:** Nemegt Formation. Well-watered but arid near-desert environment dominated by low scrub, lakes, and dry woodland. **Size:** WS >35cm (1.1ft); BL 1m (3.3ft); TL unknown **Features:** Strange & highly advanced basal caenagnathiformes. Head small w/ conspicuous dome above eyes. Beak high & prominent, w/ serrated edges. Neck slender. Wings very small but well-developed. Flat, sharp ridge on ulna probably anchored significant display feathers when unfolded. Legs extremely long & slender. **Biology:** Bonebeds & trackways indicate this species was very common in the Nemegt marshlands & highly gregarious, congregating near lakes & ponds in huge flocks. May have dabbled for waterplants & aquatic invertebrates.

Caenagnathoids

The caenagnathoids include two primary divisions, the oviraptorids and caenagnathids. Oviraptorids, or "egg seizers," so named for the mistaken belief that they raided nests which later turned out to be their own. These advanced caenagnathiform desert birds had toothless beaks and, in most cases, very long necks. The wings were generally large, and in the advanced "ingeniines", the wing bones were highly reduced but probably supported large remiges. These broad wings were used primarily in brooding and shading nests in desert environments. Nesting was communal, with multiple females laying distinctively long, narrow eggs into shared clutches, which were then incubated by the males. Nesting specimens are common, and indicate that incubation lasted for long periods of time, probably for upwards of 40 consecutive days, similar to modern flightless desert birds. Oviraptorids seem to have preferred nesting sites in the soft soil near seasonal streams in their otherwise arid, desert environments. Very few species are known from environments other than high desert, and they seem to have preferred hot, arid environments with little running water and sparse plant life. They were probably generalists, taking both small game and seeds or other tough plant material.

Many oviraptorids bore prominent casques, though this was dependent on size. It is likely that casques only developed in large adults, though some species (like *Conchoraptor gracilis*) were uniformly small and may never have developed casques even when fully mature. It is possible that some species comprised entirely of specimens lacking casques will turn out to be juveniles and/or females of casqued species.

Possibly a sub-group of the oviraptorids, or alternately a more primitive caenagnathiform lineage, caenagnathids are mostly known from very fragmentary remains which make it difficult to compare closely related species. The only species known from near-complete remains is the unnamed, giant Lancian species. Some skeletal features appear more primitive than other caenagnathiformes, which may represent the kind of reversals typical of advanced large flightless birds.

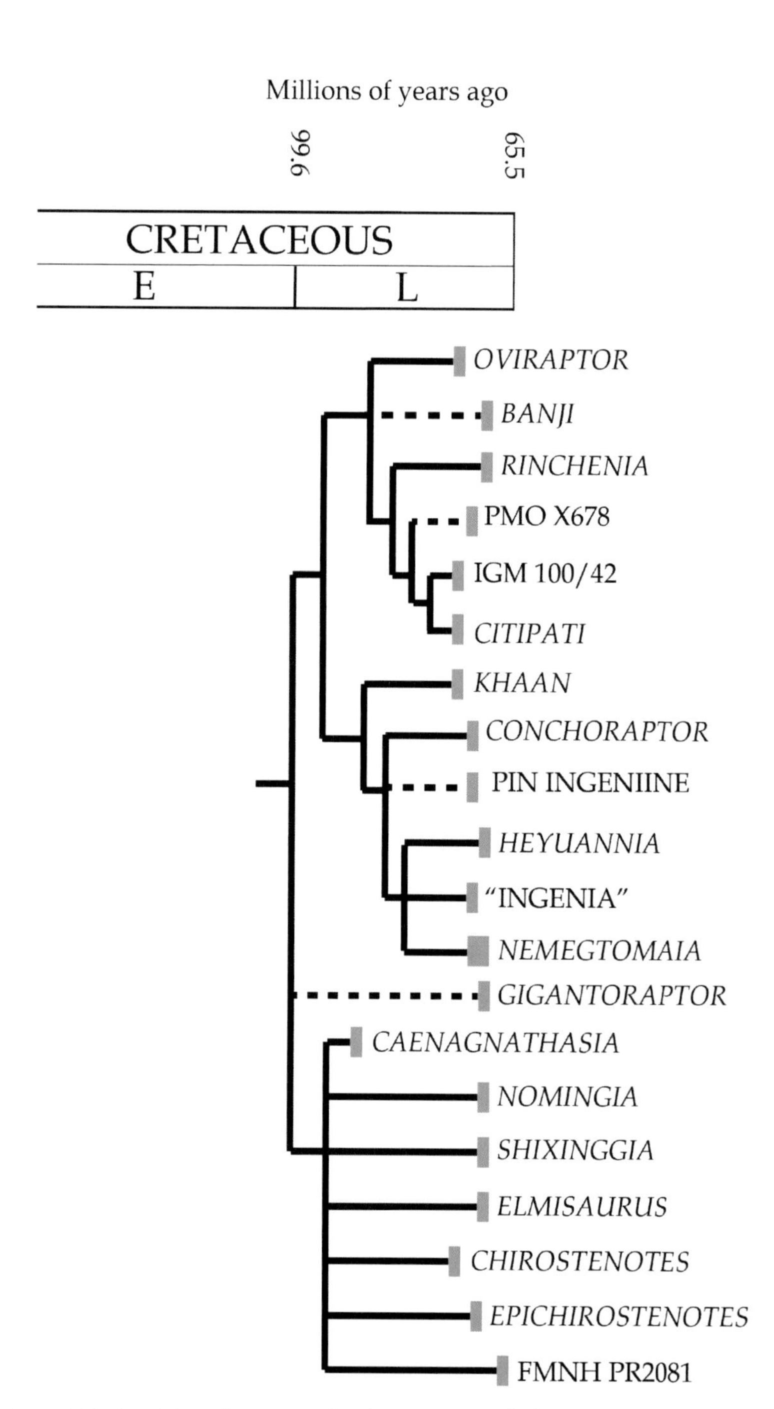

Above: Relationships of caenagnathoids over time. Phylogeny approximated based on Senter 2007 and Longrich et al. 2010.

Egg Seizer Fond of Ceratopsians ***Oviraptor philoceratops***
Time: 75 Ma ago **Location:** Omnogovi, Mongolia
Habitat: Bayan Dzak, Djadochta Formation. High desert. Dune fields and arid scrubland. **Size:** WS >1m (3.3ft); BL ~1.4m (4.6ft); TL unknown **Features:** Large but somewhat primitive oviraptorids. Head relatively long. Casque shape unknown but possibly shorter than related species. Wings large w/ long digits & large claws. **Biology:** Partially carnivorous, known to have taken small desert lizards. Name refers to the initial inference that they preyed on nests of the contemporary *Protoceratops andrewsi*, but which later proved to be their own.

Zamyn Khondt Egg Seizer **No scientific name (specimen IGM 100/42)**
Time: 75 Ma ago **Location:** Omnogovi, Mongolia **Habitat:** Zamyn Khondt, Djadochta Formation (see above) **Size:** WS >1.3m (4.4ft); BL ~2.1m (7ft); TL unknown **Features:** Large oviraptorids. Head square in profile w/ narrow bill transitioning into a large, square casque above the nostrils. Neck long. Wings large w/ long, clawed digits. Legs relatively short. **Biology:** Sometimes considered a growth stage of *C. osmolskae*, the only known fossil of this species comes from different & possibly slightly younger deposits. It may therefore be a descendant species instead.

Osmolska's Citipati ***Citipati osmolskae***
Time: 75 Ma ago **Location:** Omnogovi, Mongolia **Habitat:** Ukhaa Tolgod, Djadochta Formation (see above). **Size:** WS >1.5m (5ft); BL ~2.7m (9ft); TL unknown **Features:** Large species with a small, round head. Casque mid-sized & slightly pointed. Edge of upper bill serrated. Neck long, wings large w/ long, powerfully clawed digits. Major digit longer than lower wing bones (ulna). **Biology:** Nested communally in dune fields. Dug bowl-shaped nests in the sand, sometimes in close proximity to those of smaller species including *Byronosaurus jaffei*. Males incubated large clutches of long, narrow-oval eggs, which females deposited in circular patterns.

Mongolian Rinchen ***Rinchenia mongoliensis***
Time: 70 Ma ago **Location:** Omnogovi, Mongolia **Habitat:** Nemegt Formation. Well-watered but arid, near-desert environment dominated by low scrub, lakes, and dry woodland. **Size:** WS >90cm (3ft); BL 1.7m (5.6ft); TL unknown **Features:** Advanced oviraptorids w/ very large, oval casque. Wings large w/ long digits similar to *Citipati osmolskae*. **Biology:** Possibly direct descendants of the earlier long-handed oviraptorids of the genus *Citipati*.

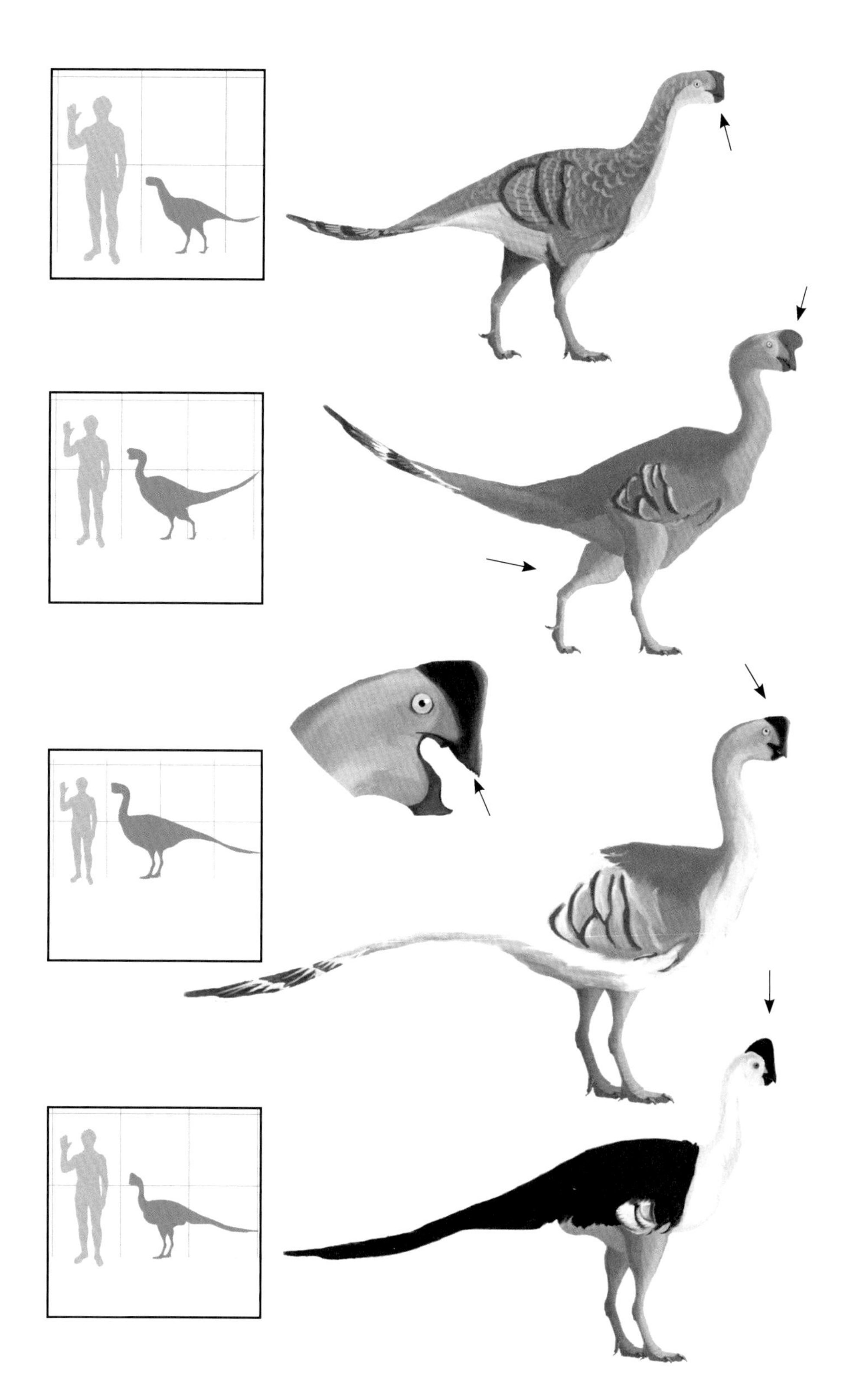

Mitre-crested Egg Seizer **No scientific name (specimen number PMO X678)**
Time: 72 Ma ago **Location:** Omnogovi, Mongolia **Habitat:** Hermin Tsav, Barun Goyot Formation (see above). **Size:** WS unknown; BL ~1.5m (5ft); TL unknown **Features:** Similar to, but smaller than, *Citipati osmolskae*, w/ relatively larger casque which was tall, narrow, pointed, & curved slightly forward. Wings small w/ short, even-length digits similar to *Conchoraptor gracilis*. **Biology:** Potentially mature growth stage of *C. gracilis* (Paul 2010).

Slender Shell Thief ***Conchoraptor gracilis***
Time: 72 Ma ago **Location:** Omnogovi, Mongolia **Habitat:** Hermin Tsav, Barun Goyot Formation. High desert. Dune fields and arid scrubland. **Size:** WS unknown; BL ~1.4m (4.6ft); TL unknown **Features:** Head round w/ heavy beak forming a small crest above the nostrils. Neck long & slender. Body relatively heavy compared to "Ingenia". Wing digits longer & more slender than "Ingenia". Tail short but very deep. **Biology:** These have been suggested to represent juvenile specimens of related species, but because all known specimens are small in size, it is likely that they are an advanced lineage which retained juvenile characteristics (small size, lack of casque) into adulthood.

Big-beaked Shell Thief **No scientific name**
Time: 72 Ma ago **Location:** Omnogovi, Mongolia **Habitat:** Hermin Tsav, Barun Goyot Formation (see above) **Size:** WS unknown; BL ~1.4m (4.6ft); TL unknown **Features:** Similar to *Conchoraptor gracilis* but with a particularly heavy beak, sporting two 'prongs' on either side of the pointed bill.

McKenna's Lord ***Khaan mckennai***
Time: 75 Ma ago **Location:** Omnogovi, Mongolia **Habitat:** Ukhaa Tolgod, Djadochta Formation (see above). **Size:** WS >60cm (2ft); BL 1.3m (4.2ft); TL unknown
Features: Small, round head w/ no crest. Neck very long & slender. Wings small w/ long digits. Alular digit long & slender, major & minor digits slightly longer. **Biology:** Possibly the juvenile form of *Citipati osmolskae* (Paul 2010). While the orientation of the skull bones differs between these two species, this may be attributable to the formation of the casque in adults. The difference in wing anatomy (major & minor digits short & slender vs. long & robust) is easily explained by overall growth & presumably the enlargement of the remiges with age.

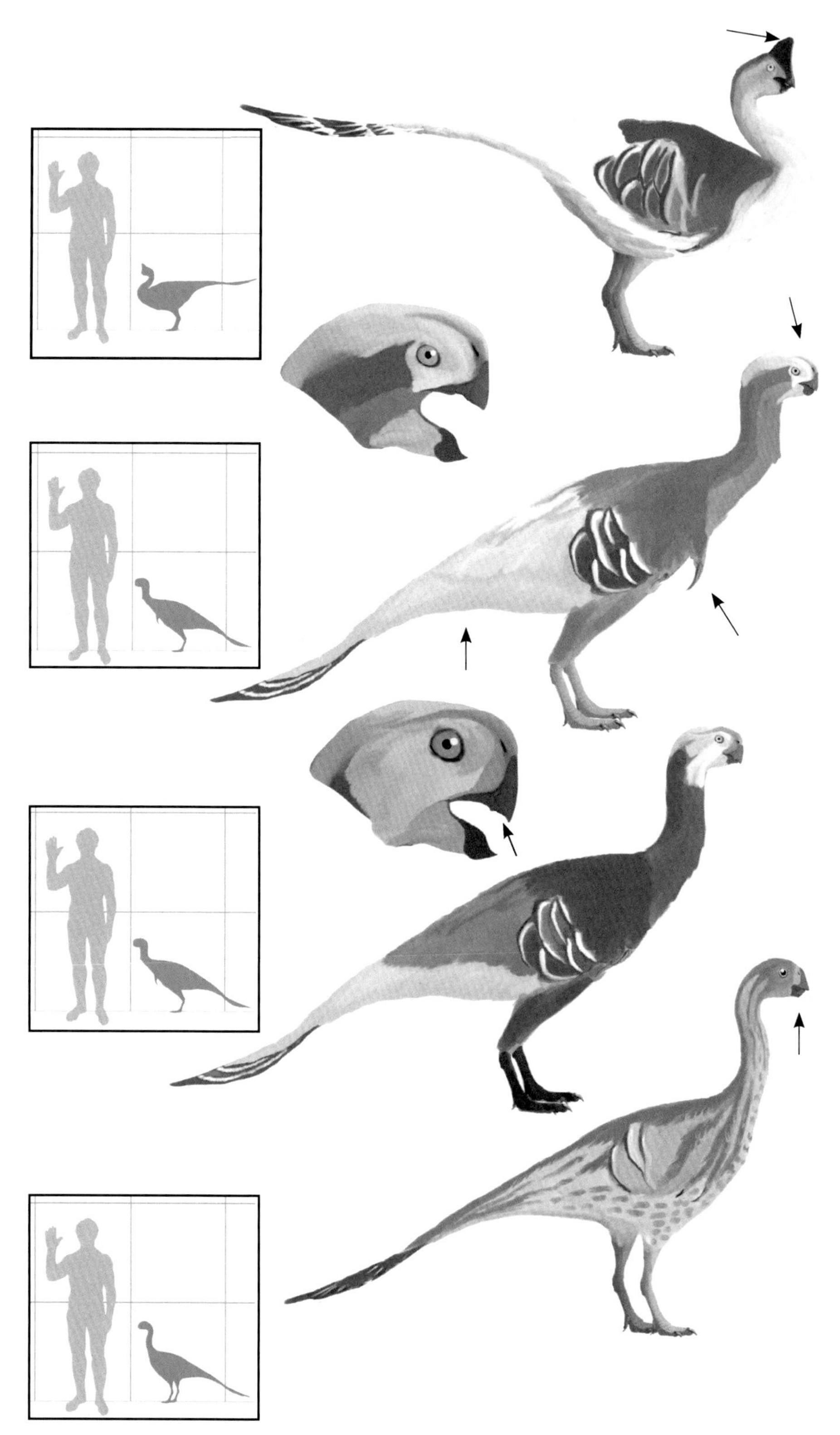

Huang's Heyuan ***Heyuannia huangi***
Time: Uncertain (Maastrichtian?) **Location:** Guangdong, China **Habitat:** Zhutian Formation **Size:** WS >50cm (1.6ft); BL 1.5m (4.9ft); TL unknown **Features:** Head small w/ large eyes & short beak. Crest absent. Neck long & slender. Wings small, w/ short, partially fused wing bones.

Yanshin's Ingeni **"Ingenia"** ***yanshini***
Time: 72 Ma ago **Location:** Omnogovi, Mongolia **Habitat:** Hermin Tsav, Barun Goyot Formation (see above). **Size:** WS >80cm (2.6ft); BL ~1.6m (5.2ft); TL unknown **Features:** Head small & slightly flat & elongated w/ no crest. Wings small w/ very small digits & large claws. **Biology:** The name "Ingenia" is preoccupied & will be replaced.

Barsbold's Nemegt Mother ***Nemegtomaia barsboldi***
Time: 72-70 Ma ago **Location:** Omnogovi, Mongolia **Habitat:** Barun Goyot and Nemegt Formations (see above) **Size:** WS >60cm (1.9ft); BL 1.8m (6ft); TL unknown **Features:** Similar to "Ingenia" *yanshini* but w/ a large, low casque similar to that of *Citipati osmolskae*. Wings small, w/ short, stout digits. May represent mature individuals of "Ingenia" *yanshini*.

Stripe-crested Dragon ***Banji long***
Time: Uncertain (Campanian or Maastrichtian) **Location:** Jiangxi, China **Habitat:** Nanxiong Formation. **Size:** WS unknown; BL ~55cm (1.8ft); TL unknown **Features:** Crest prominent & raised above the head sharply from the back. Deep striations on the underlying bone indicate large keratin component in life, probably for display, possibly ridged. **Biology:** The smallest known oviraptorid species, though the only known specimen may actually be a juvenile.

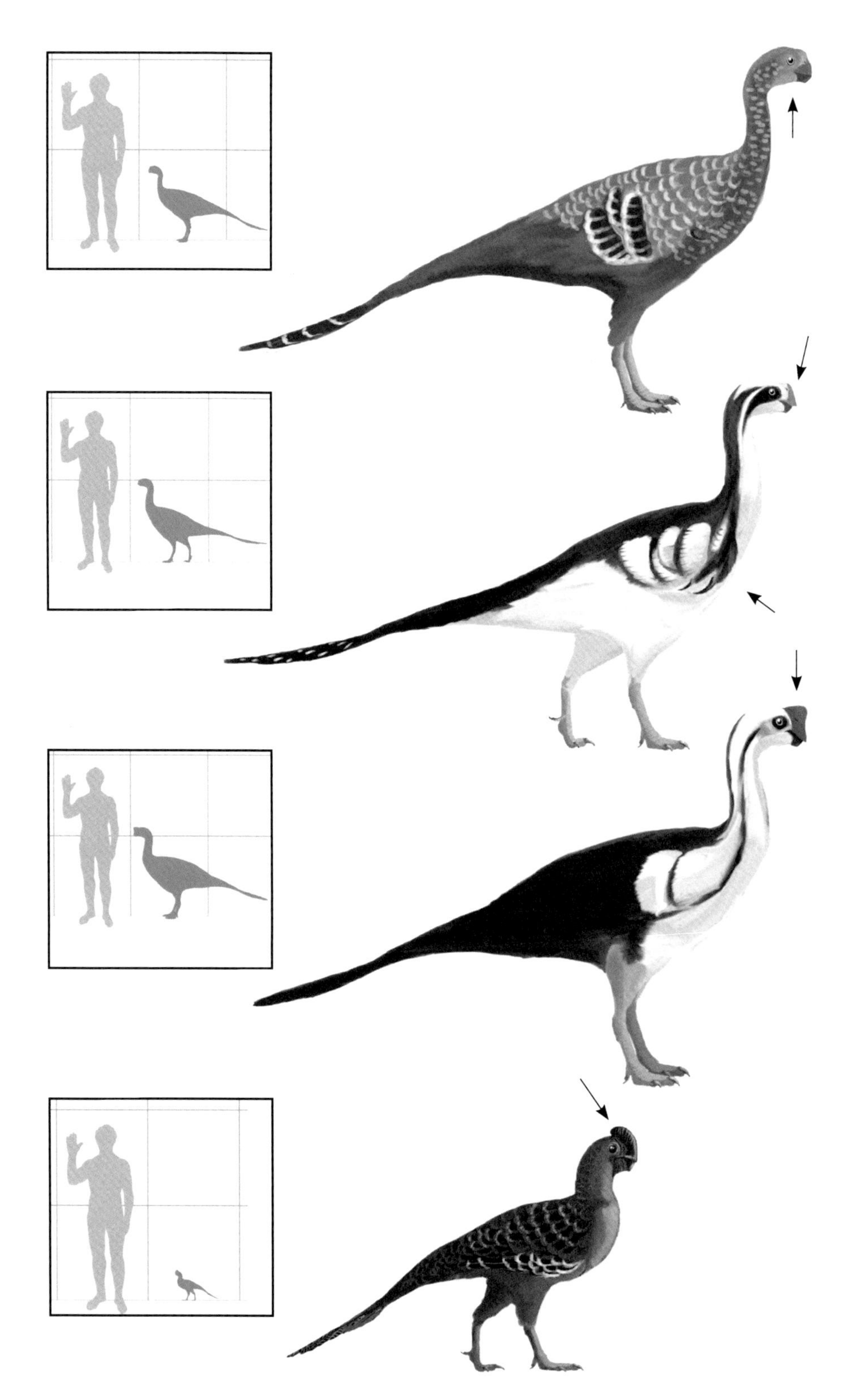

Erlian Gigantic Raptor ***Gigantoraptor erlianensis***
Time: 70 Ma ago **Location:** Omnogovi, Mongolia **Habitat:** Nemegt Formation (see above). **Size:** WS >4m (13ft); BL ~7m (23ft); TL unknown **Features:** Head relatively small w/ heavy, beaked jaws. Wings large w/ long digits. Legs long & powerful. **Biology:** One of the largest winged dinosaur species known. Several large, elongate oval eggs w/ oviraptorid embryos from the same formation may belong to this species.

Martinson's Asian Recent Jaw ***Caenagnathasia martinsoni***
Time: 90 Ma ago **Location:** Uzbekistan **Habitat:** Bissekty Formation **Size:** WS unknown; BL ~55cm (1.8ft); TL unknown **Features:** Small caenagnathiforms known only from the lower jaws, which bore blunt, rounded & beaked tip. Beak relatively smooth compared with larger relatives. **Biology:** Likely omnivorous as w/ other caenagnathiformes, but the smooth bill edge may imply a more herbivorous diet than most.

Gobi Noming ***Nomingia gobiensis***
Time: 70 Ma ago Location: Omnogovi, Mongolia **Habitat:** Nemegt Formation. Well-watered but arid near-desert environment dominated by low scrub, lakes, and dry woodland. **Size:** WS unknown; BL ~1.8m (6ft); TL unknown **Features:** Small caenagnathids known from hind limbs, pelvis & tail. Legs mid-length to short but powerful. Tail short as in other caenagnathoids but w/ last few vertebrae fused into a pygostyle-like structure. **Biology:** The distinctive pygostyle-like tail fusion is not found in some other caenagnathiformes & may imply an unusually large feather frond on the tail.

Forgotten Shixing ***Shixinggia oblita***
Time: 70 Ma ago **Location:** Guandong, China **Habitat:** Pingling Formation **Size:** WS unknown; BL ~1.8m (6ft); TL unknown **Features:** Known from a partial skeleton missing the head and neck, and characterized by internal anatomy including increased presence of air sacs in the legs. **Biology:** These show characteristics typical of both caenagnathids & oviraptorids, & possibly represent relatively primitive members of the group.

Rare Hand Lizard ***Elmisaurus rarus***
Time: 70 Ma ago **Location:** Omnogovi, Mongolia **Habitat:** Nemegt Formation (see above) **Size:** WS unknown; BL ~1.8m (6ft); TL unknown **Features:** Known from wings, feet, & possibly a partial skull. Distinguished from *Chirostenotes pergracilis* by the presence of unique internal anatomy including a vascular opening in the tarsus & a prominent muscle attachment point at the ankle. **Biology:** Probably similar to *Chirostenotes* and/or *Avimimus* in terms of ecology. This was the first caenagnathid species known from both wing & foot bones, allowing the various pieces of *Chirostenotes* (previously attributed to different animals) to be brought together.

Thin Narrow Hand ***Chirostenotes pergracilis***
Time: 75 Ma ago **Location:** Alberta, Canada **Habitat:** Dinosaur Park Formation, seasonally arid lowland plains dominated by braided river systems and small forests. **Size:** WS unknown; BL ~1.5m (5ft); TL unknown **Features:** Known from isolated jaw, hands & hind legs, but very similar to (though significantly smaller than) the later Lancian caenagnathids. Lower jaw wide but narrow when viewed from the side, w/ long, slightly up-curved beak. Upper jaw unknown, but likely similar to the Lancian species, w/ broad, somewhat spoon-shaped bill. Alular digit short & stout w/ large, curved claw. Major & minor digits long & relatively slender. Internally characterized by a middle metatarsal with a diamond-shaped cross-section. **Biology:** As with some other caenagnathids, two varieties are known for this species: a slender morph and a robust morph. The slender morph has been given its own name, *Chirostenotes elegans*, but the difference is probably due to sex.

Currie's Higher Narrow Hand ***Epichierostenotes curreii***
Time: 72 Ma ago **Location:** Alberta, Canada **Habitat:** Horseshoe Canyon Formation, seasonally arid lowland plains dominated by braided river systems and small forests. **Size:** WS unknown; BL ~1.7m (5.7ft); TL unknown **Features:** Known from a partial skeleton, very similar to *Chirostenotes pergracilis*, differing only in size and time period, as far as it's possible to tell. **Biology:** Probably a direct descendant species of *C. pergracilis*.

Lancian Recent Jaw **No scientific name (specimen number FMNH PR2081)**
Time: 65.5 Ma ago **Location:** South Dakota, USA **Habitat:** Hell Creek Formation. Forested near-coastal flood plains dominated by flowering shrub species and coniferous trees. **Size:** WS >2m (6.6ft); BL 3.7m (12ft); TL unknown **Features:** Very large. Head long w/ prominent, large rounded casque. Beak straight but broad & rounded. Neck relatively long, though not as long as some oviraptorids. Wings large & clawed. Legs long, but feet relatively small. Tail short but very deep, with fused vertebrae at the tip. **Biology:** The fused, pygostyle-like tail vertebrae may imply a large tail fan or may simply be the byproduct of tail shortening. The broad, flat bill w/ straight edges suggests a mostly herbivorous diet, though some degree of omnivory is likely, as in other caenagnathiformes. *Chirostenotes elegans* specimens have been reported from the same formation, despite the ten million years chronological gap. As with *C. pergracilis*, these "*C. elegans*" specimens probably represent the smaller, more slender females of this species.

Basal Eumaniraptorans & Deinonychosaurians

The deinonychosaurians, or "terrible claw lizards", currently represent one of the earliest and most primitive known lineages of frond-tailed birds (along with their sister lineage, the *Avialae*). The fossil record of early deinonychosaurians is relatively complete, and primitive members are known from good fossil remains and feather impressions. Not surprisingly, primitive members of each group are very similar to each other, and to primitive avialans, making it clear that the deinonychosaurian lineage and the one leading to modern birds evolved from a common ancestor very much like *Archaeopteryx lithogrpahica* or *Xiaotingia zhengi*.

Primitive deinonychosaurs appear to have been glissant (i.e. capable of gliding flight), with some taking steps toward powered, flapping flight. *Microraptor zhaoianus* had well-developed wings, and possessed an additional set of "hind wings" formed from vaned feathers on the lower legs and feet, which would have formed a biplane-like configuration when gliding or parachuting from trees. In *Archaeopteryx*, *Microraptor*, and *Rahonavis*, the wings were large enough and powerful enough to have allowed clumsy, level bursts of flight, though gliding was probably the preferred mode of aerial transport.

In addition to the early small, glissant species, deinonychosaurians exhibited a common trend found among even modern birds: that of flightless ground birds becoming larger and more cursorial (i.e. adapted to a ground-dwelling lifestyle). These later, larger species (members of the group *Eudromaeosauria*) include the famous "raptors," which did not actually resemble the reptilian monsters depicted in popular cinema, but were very large predatory ground birds, some with wings of substantial size (evidence for this comes from feather anchor points found on the wing bones of *Velociraptor mongoliensis*). The "raptors" did not closely resemble their carnosaurian cousins, but rather oversized *Archaeopteryx*. As paleontologist Mark Norell stated in an interview on the subject:

"The more that we learn about these animals the more we find that there is basically no difference between birds and their closely related dinosaur ancestors like *Velociraptor*. Both have wishbones, brooded their nests, possess hollow bones, and were covered in feathers. If animals like *Velociraptor* were alive today our first impression would be that they were just very unusual looking birds."

Even the characteristically enlarged "sickle claw" on the second toe of most deinonychosaurians probably did not begin as a weapon to hold and kill prey. Rather, their shape (and the shape of the forelimb claws) more closely matches the claws of climbing animals. The sickle claws and wing claws alike were also attached in a way ideal for the transfer of stress loads to the rest of the foot or wing, unlike the expected anatomy of a slicing weapon. It is likely that these claws were first used for climbing trees (like the crampons used by utility pole linemen) in small, glissant species, and were later adapted for prey capture in their larger, ground-dwelling descendants.

This trend from small gliders to large ground birds is not unique to the deinonychosaurians, but their reversion to a predatory lifestyle may be. Many lineages of early maniraptorans (birds and their closest relatives) appear to have been omnivorous, herbivorous or insectivorous, and only among the eudromaeosaurians did hypercarnivory (diets including mainly large vertebrate prey items) evolve.

While deinonychosaurians must have emerged at least in the Middle Jurassic, all primitive known members of this family are Late Jurassic or younger in age, and may have departed significantly from the ancestral body plan, inferred to have been small and large-winged.

The first known Mesozoic bird, *Archaeopteryx lithographica*, was found in 1861 and named based on a single feather. The name was transferred to a more complete skeleton in 2011. Several species of archaeopterygid have been recognized in the past from the Solnhoffen limestone of Bavaria. However, most of those are probably specimens of *A. lithographica* at various stages of growth. *Archaeopteryx* is traditionally considered to be more closely related to modern birds than to deinonychosaurians, though it is possible that archaeopterygids are on the deinoychosaurian line, or basal to the avialan/deinonychosaurian split.

Huxley's Near Bird ***Anchiornis huxleyi***
Time: 160 Ma ago **Location:** Liaoning, China **Habitat:** Tiaojishan Formation
Size: WS 37cm (1.2ft); BL 40cm (1.3ft); TL 45cm (1.5ft) **Features:** Head triangular with bluntly pointed snout & prominent red/black crown. Snout feathered to near the tip. Contour feathers dull black w/ red speckles on the face. Wings relatively short & rounded with narrow, symmetrical remiges. Remiges & coverts densely layered, w/ coverts extending to near the remix tips (Longrich & al. 2012). Fingers completely feathered. Remiges & coverts white w/ black spangles in even rows on primaries & uneven spots on secondaries. Legs long w/ short hind wings. Hindwing feathers spangled similarly to forewings, long near foot & tapering towards body. Large but weakly curved sickle claw. Toes completely feathered. Tail very long w/ spangled rectrices extending to the base. **Biology:** Life coloration revealed through study of preserved melanin, but different coloration may have existed between sexes, growth stages or populations. Rounded wings w/ unspecialized remiges indicate a loss of aerial ability, but parachuting or limited gliding may have been possible. Striking coloration of the crown & wing feathers indicates a role in display.

Daohugou Foot Feather ***Pedopenna daohugouensis***
Time: 155 Ma ago **Location:** Inner Mongolia, China **Habitat:** Daohugou Formation
Size: WS unknown; BL ~80cm (2.6ft); TL unknown **Features:** Known only from the foot/lower leg. Legs long. Hind wings prominent but reduced, w/ primary feathers much longer than secondaries. Hallux very slender & not reversed. **Biology:** Like *Anchiornis*, the foot feathers were relatively short & not strongly vaned, so would not have imparted much aerodynamic assistance.

Zheng's Xiaoting ***Xiaotingia zhengi***
Time: 160 Ma ago **Location:** Liaoning, China **Habitat:** Tiaojishan Formation **Size:** WS ~60cm; BL ~70cm; TL unknown **Features:** Head triangular w/ elongated snout. Wings large. Legs long & bearing hind wings. Fingers & toes completely feathered. Sickle-claw on second toe. **Biology:** Feather details unknown, but restored as intermediate between *A. huxleyi* and *A. lithographica*.

Lithographic Ancient Wing ***Archaeopteryx lithographica***
Time: 150 Ma ago **Location:** Bavaria, Germany **Habitat:** Solnhofen Formation. Arid tropical island in Tethys sea. Low scrub and beaches surrounding sheltered lagoons. **Size:** WS 77cm (2.5 ft); BL 55cm (1.8); TL 60cm (2ft) **Features:** Wings broad & rounded, with aerodynamic feathers, at least some coverts black in color. Coverts very long, with secondary covers reaching nearly to the tips of secondary remiges (Longrich & al. 2012). Proportionately short portion of clawed digits emergent. Legs short & covered in long feathers which extend past the ankle. Tail long w/ large feather frond extending to the base. Rectrices longer towards the tip, & forming a somewhat split 'V' shaped tail. **Biology:** Primitive wing anatomy & relatively weak primary feather rachides indicate flightlessness. Gliding possible, but these were more likely beachcombers. Most specimens small (pigeon size or less), & some specimens have been classified in different species (including *Archaeopteryx siemensi, Wellnhoferia grandis,* & *Jurapteryx recurva*), though these all likely represent growth stages of a single species. Wing & feather proportions remained relatively constant throughout growth.

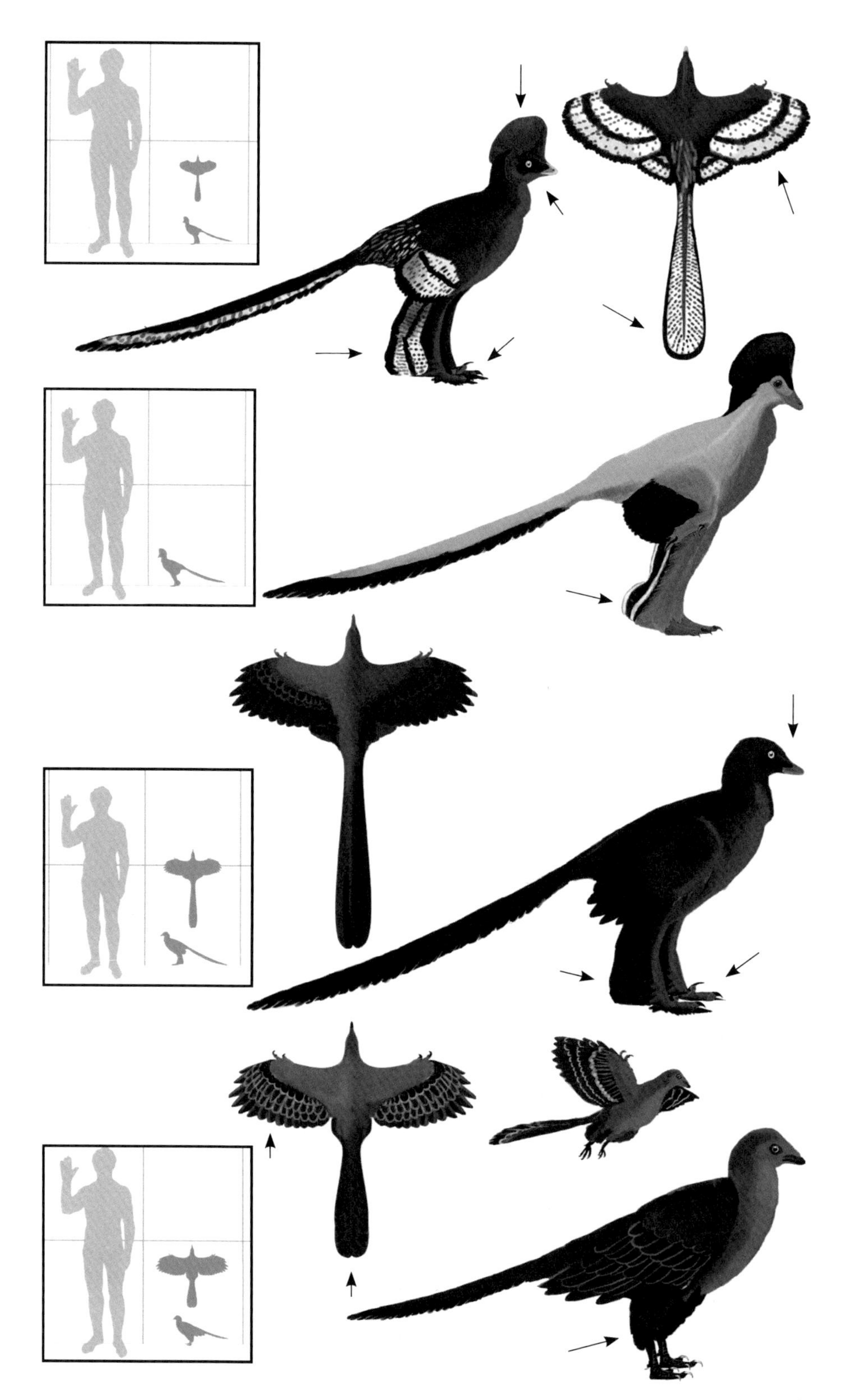

Microraptorians

Microraptorians were the smallest and among the most primitive ornithodesmids, most retaining many features in common with archaeopterygids and early troodontids. The earliest members of this group were large compared to other primitive birds, though later members independently re-acquired small sizes comparable to archaeopterygids. None were as large as the moderately sized members of the other ornithodesmid groups.

Microraptorians appear to have been more predatory than most other early birds, beginning a trend that would reach its apogee in the large eudromaeosaurians. Most had long, narrow snouts with recurved and partially serrated teeth. The claw borne on digit II was larger than in archaeopterygids and many troodontids of the same size, though it was still relatively broad compared to eudromaeosaurians and may have been primarily a climbing tool. There is evidence that even the smallest species, such as *Microraptor zhaoianus*, occasionally took prey that approached or exceeded their own body size.

Suggestions that some microraptorians were venomous, and that they had long, protruding, fang-like teeth, are incorrect, and were based on misinterpretation of teeth that had come out of their sockets during fossilization.

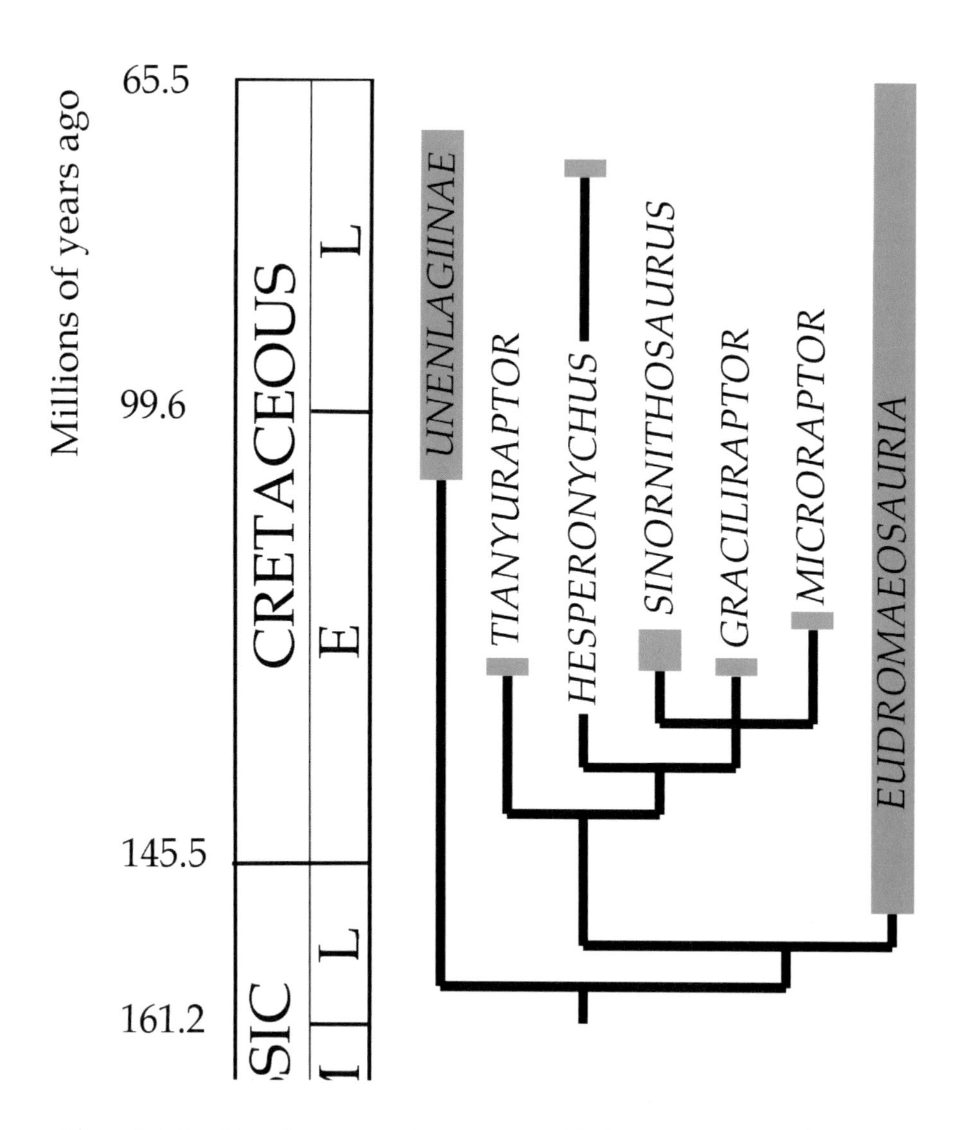

Above: Relationships of microraptorians over time. Phylogeny approximated based on Senter et al. 2012.

Ostrom's Tianyu Robber ***Tianyuraptor ostromi***
Time: 122 Ma ago **Location:** Liaoning, China **Habitat:** Upper Yixian Formation (see above) **Size:** WS >60cm (2ft); BL 1.7m (5.5ft); TL unknown **Features:** Snout broad. Neck short & covered in very long feathers. Wings small & degenerate. Legs very long. Tail very long & slender. **Biology:** Highly reduced wishbone & long legs indicate a fully terrestrial lifestyle.

Hanqing's Small Raptor ***Microraptor hanqingi***
Time: 122-120 Ma ago **Location:** Liaoning, China **Habitat:** Jiufotang Formation, temperate-subtropical swampland dominated by ginkgo and conifer trees, set among shallow lakes and stagnant waterways. **Size:** WS 75cm (2.5ft); BL 85cm (2.8ft); TL 90cm (3ft) **Features:** Very similar to *M. zhaoianus* (below), but much larger. Feathers poorly preserved. Tarsal remiges very short compared to *M. zhaoianus* & *C. pauli* specimens (~equal to tarsal length). Possible growth stage of *M. zhaoianus*; if so, would imply reduction in aerial ability even after sexual maturity was reached.

Zhao's Small Raptor ***Microraptor zhaoianus***
Time: 122-120 Ma ago **Location:** Liaoning, China **Habitat:** Yixian (see above) & Jiufotang Formation, temperate-subtropical swampland dominated by ginkgo and conifer trees, set among shallow lakes and stagnant waterways. **Size:** WS 57cm (1.8ft); BL 60cm (2ft); TL 68cm (2.2ft) **Features:** Head large. Snout narrow w/ fine teeth. Neck short. Primaries long & aerodynamic, w/ slightly backward-curved shafts. Secondaries short, making wings long & pointed w/ a high aspect ratio. Legs long, w/ large, rounded hindwings. Hindwing primaries & secondaries subequal in length. Tail very long & thin, w/ lozenge-shaped rectrical frond covering the final half of its length. The final two rectrices were elongated in male specimens. Plumage uniformly black with glossy iridescence. **Biology:** May have scaled trees by hugging the trunk with clawed wings to launch from a height. Rectrical frond ended in two elongated rectrices similar to early avialans, probably used in mating displays. Hunted small enantiornitheans, but also took larger prey.

Paul's Hidden Flyer ***Cryptovolans pauli***
Time: 75 Ma ago **Location:** Alberta, Canada **Habitat:** Dinosaur Park Formation. Seasonally arid lowland plains w/ braided river systems and small forests. **Size:** WS unknown; BL ~1m (3.3ft); TL unknown **Features:** Head relatively small, but snout unknown. Wings well developed, long, & pointed. Breastbone fused. Legs long, accommodating very large hindwings. Hindwing primaries & secondaries subequal in length, coverts very long compared to corresponding remiges. Hindwings extended from bases of the toes to the thighs, but decreased in size beginning at the ankles. Tail long w/ large, lozenge-shaped rectrical frond covering the final half of its length. **Biology:** Likely glided or flew weakly, w/ hind wings beneath fore-wings in a biplane configuration. Presence of fused breastbone may indicate powered flight ability. Hindwings not as aerodynamic as forewings, but may have been extended one at a time as rudders or brakes (Hall & al. 2012). Large tail frond likely used in pitch control (Habib & al. 2012). Reported presence of a feather crest inaccurate, based on misinterpretation of crushed head and neck feathers. *Microraptor gui* is likely a junior synonym, sharing the fused sternum.

Lujiatun Graceful Thief ***Graciliraptor lujiatunensis***
Time: 124.5 Ma ago **Location:** Liaoning, China **Habitat:** Lower Yixian Formation, temperate conifer and ginkgo forest set among a series of lakes fed by streams and runoff from volcanic mountains. **Size:** WS ~1.2m (4ft); BL ~1.1m (3.6ft); TL unknown **Features:** Mid-sized. Wings likely long & broad. Legs very long, possibly supporting mid- to large- sized hindwings. **Biology:** While as large as contemporary *Sinornithosaurus*, the long legs & slender proportions approached the gliding microraptorians, & they may have therefore been slightly more arboreal than the sinornithosaurs.

Millennial Chinese Bird Lizard ***Sinornithosaurus millennii***
Time: 124.5 Ma ago **Location:** Liaoning, China **Habitat:** Upper & Lower Yixian Formation (see above) **Size:** WS ~1m (3.3ft); BL ~1.2m (3.9ft); TL unknown **Size:** (juvenile specimen NGMC 91) WS 60cm (2ft); BL 65cm (2.1ft); TL ~70cm (2.3ft) **Features:** Snout long & narrow in adults, higher & more triangular in jueniles. Feathers extend to near snout-tip. Wings degenerate, w/ symmetrical remiges shorter than second digit. Legs long w/ long vaned feathers on at least the femur. Feet scaly w/ large talons & sickle claw. Tarsal feathers unknown. Tail very long in adult & short in juvenile, w/ vaned feathers extending to base. Rectrices longer near the tail tip, forming lozenge-shaped frond. Heavily mottled coloration consisting of dark brown or black & light brown to russet feathers. **Biology:** Similar to *Graciliraptor* but w/ shorter legs & smaller wings. Small wings w/ symmetrical vaned feathers indicative of flightlessness. Probably relied on the intricate banding pattern of the feathers for display/camouflage. Originally named *S. millenii*, but under the ICZN, this must be emended to *S. millennii*.

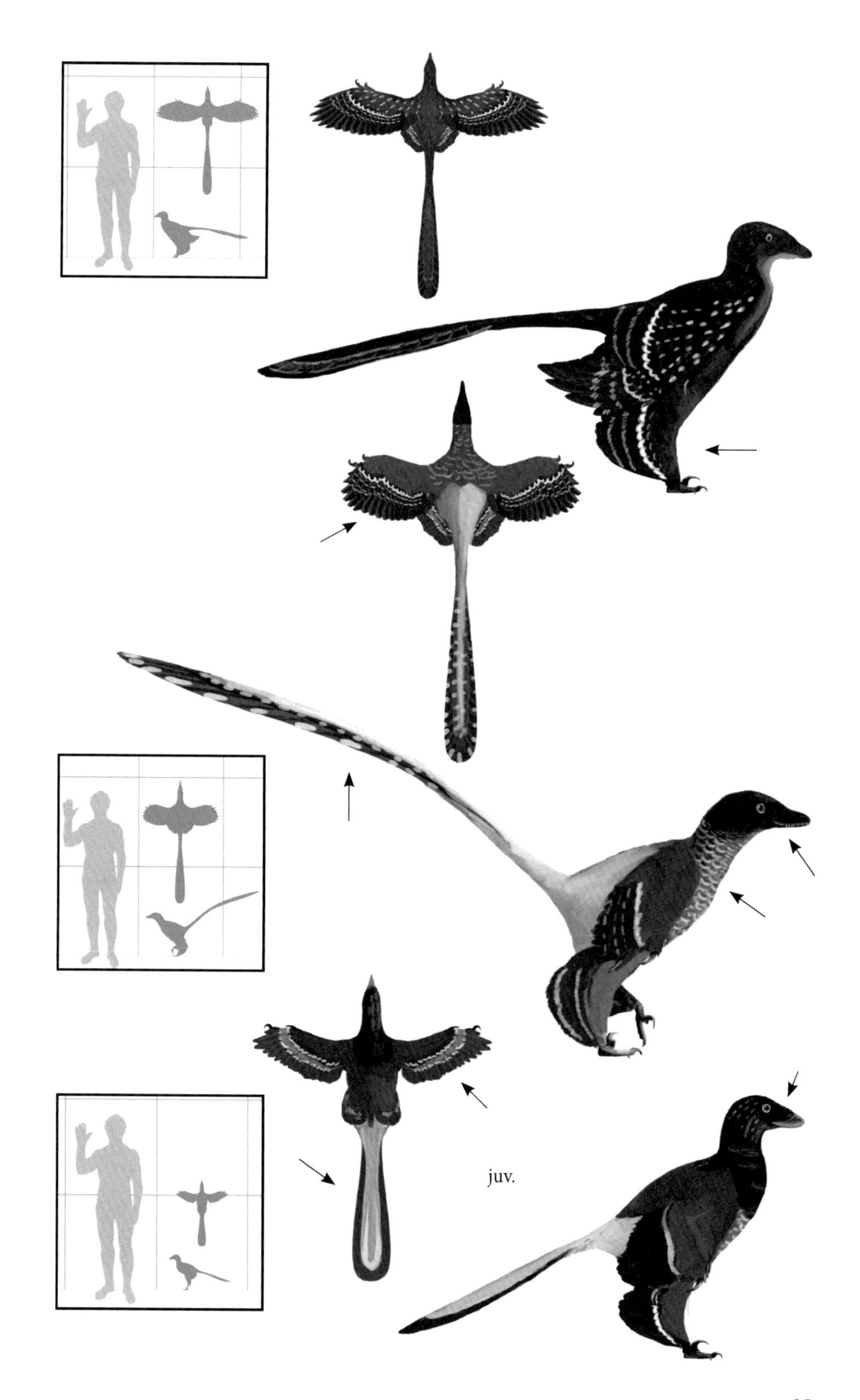
juv.

Eudromaeosaurians

"True dromaeosaurs", these mid-sized to giant flightless ornithodesmids are well known from Late Cretaceous deposits, but fragmentary fossils and isolated teeth show that the group originated in at least the Late Jurassic. Their lower legs were greatly reduced, limiting running speed significantly, but accommodating robust musculature to support the sickle claw, which was employed in grappling and subduing large prey. The sickle claws themselves were highly flattened and more suited to cutting and hooking than the somewhat broader claws of most other ornithodesmids. Almost all seem to have been hypercarnivores specializing in taking prey larger than themselves. Based on footprint evidence and fossil assemblages of multiple individuals, many seem to have been gregarious, travelling together in small flocks. As in modern flightless birds relatively far removed from their flying or gliding ancestors, their feathers had probably reverted to open-vaned plumes with non-interlocking barbs in many cases, and large species inhabiting hot, arid-environments may have lost feathers on some parts of their bodies for more effective temperature regulation. In modern birds, this loss of feathers helps facilitate heat shedding, but can also cause overheating. In many cases, featherless portions of the body can be partly or completely covered by the wing feathers when necessary.

Several lineages of eudromaeosaurians independently evolved large to gigantic sizes (a similar trend is seen in the evolution of the giant unenlagiines of the genus *Austroraptor* and giant itemirines like *Itemirus*). In both instances, these huge species are found during the early Cretaceous period, and may have competed with and ultimately lost out to deinodontids for dominance of the giant carnivore niches left by the waning of the allosauroids in the Northern Hemisphere.

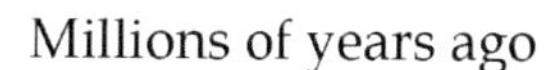

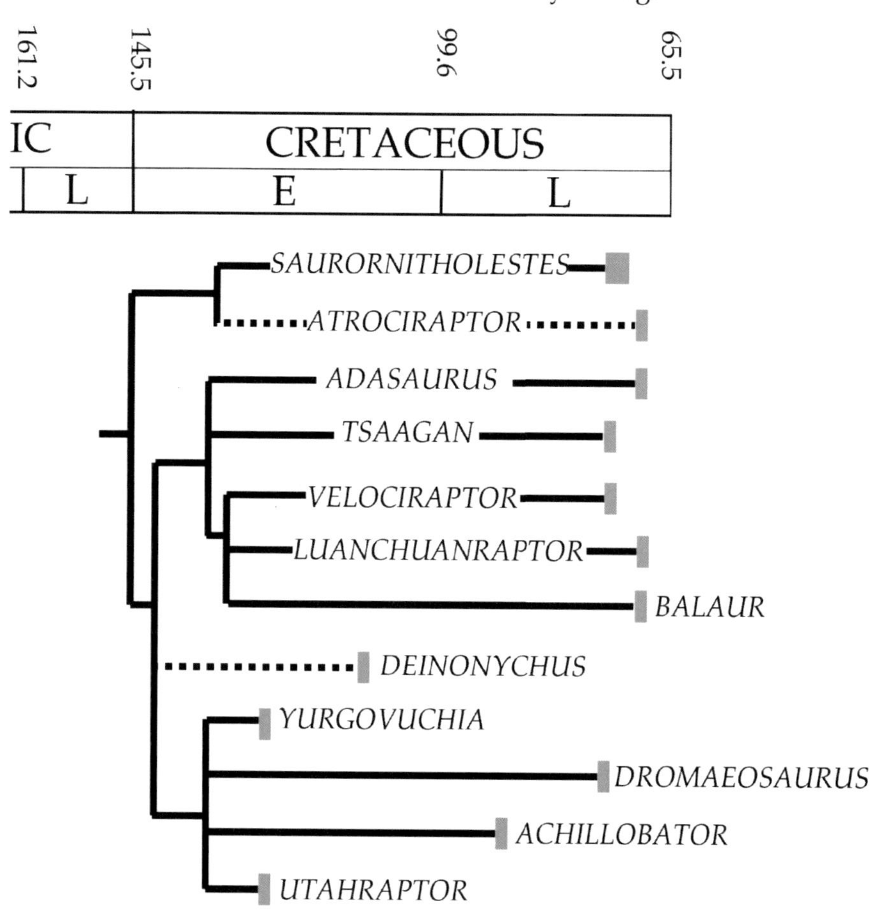

Above: Relationships of eudromaeosaurians over time. Phylogeny approximated based on Senter & al. 2012 and Turner & al. 2012

Ostrom & Mays' Utah Predator ***Utahraptor ostrommaysorum***
Time: 126 Ma ago **Location:** Utah, USA **Habitat:** Yellow Cat Member, Cedar Mountain Formation. Open, marshy mud flats. **Size:** WS unknown; BL ~6.5m (21.3ft); TL unknown **Features:** Legs short & powerful. Very large sickle claws. Tail relatively long. **Biology:** One of the largest known winged dinosaur species. Probably carnivores specializing in large prey such as ornithopods & juvenile sauropods. Extremely large size may have been attained due to the lack of large carnosaurs in the ecosystem following the extinction of allosaurids, megalosaurids, and ceratosaurids in the region, allowing eudromaeosaurians to fill niches normally occupied by other theropods. The arrival of carcharodontosaurs (*Acrocanthosaurus atokensis*) in the region corresponded with the extinction of *U. ostrommaysorum*.

Gigantic Achilles Tendon Hero ***Achillobator giganticus***
Time: 90 Ma ago **Location:** Dornogovi, Mongolia **Habitat:** Bayan Shireh Formation. Probably similar to the later Nemegt Formation, a well-watered but arid near-desert environment dominated by low scrub, lakes, and dry woodland. **Size:** WS >1.2m (4ft); BL ~5m (16.5ft); TL unknown **Features:** Snout very high & squared. Body short & deep, w/ especially long pubis. Wings small but strongly clawed. Legs very short, giving a squat appearance. Tail very long. **Biology:** May have used short but strong wings to grapple prey or buffet rivals. Legs unusually short, w/ the tibia & tarsus each shorter than femur. Probably slow-moving ambush predators.

Albertan Running Lizard ***Dromaeosaurus albertensis***
Time: 75 Ma **Location:** Alberta, Canada **Habitat:** Dinosaur Park Formation. Seasonally arid lowland plains dominated by braided river systems and small forests. **Size:** WS >unknown; BL 1.7m (5.6ft); TL unknown **Features:** Head box-like in profile, w/ long, square snout. Legs relatively long compared to contemporary *S. explanatus*. Tail more flexible than most other eudromaeosaurians. **Biology:** Longer legs & stouter jaws & teeth indicate that these may have been pursuit predators of larger prey than their contemporaries.

Marshall's Savage Robber ***Atrociraptor marshalli***
Time: 70 Ma **Location:** Alberta, Canada **Habitat:** Horseshoe Canyon Formation, seasonally arid lowland plains w/ braided river systems and small forests. **Size:** WS unknown; BL ~1.7m (5.6ft); TL unknown **Features:** Head box-like in profile, with short, rounded snout. **Biology:** Probably similar in habits to *Dromaeosaurus*. Heavy, square skull & robust teeth indicate that these took relatively large prey.

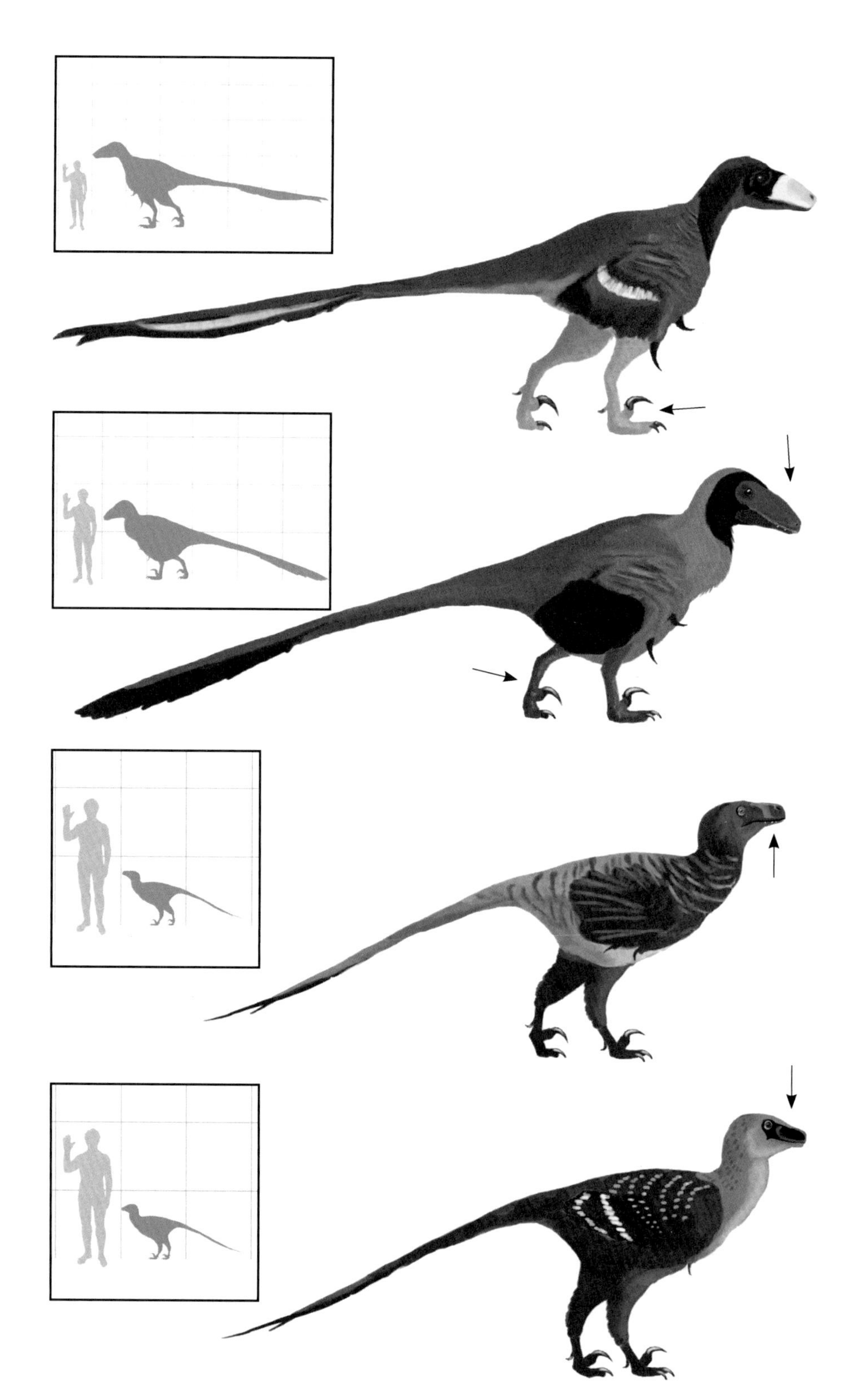

Doelling's Coyote ***Yurgovuchia doellingi***
Time: 126 Ma ago **Location:** Utah, USA **Habitat:** Lower Yellow Cat Member, Cedar Mountain Formation. Open, marshy fern prairies dominated by iguanodonts, sauropods, pseudosuchians and polocanthids. **Size:** WS unknown; BL 2.4m (7.9ft); TL unkown **Features:** Large eudromaeosaurs known from partial skeletons. Neck held relatively straight in resting posture (due to relatively smooth surface of the vertebrae). Tail more flexible than most other eudromaeosaurians. **Biology:** One of several eudromaeosaurians in this ecosystem, these were apparently the largest, nearly twice the size of an unnamed species of itemirine from the same formation. Possibly a precursor to slightly younger giant dromaeosaurines of the genus *Utahraptor*, the relatively stout neck may have helped these to tackle relatively large prey.

Counter-balanced Terrible Claw ***Deinonychus antirrhopus***
Time: 110 Ma ago **Location:** Montana, Oklahoma & Wyoming, USA **Habitat:** Cloverly & Antlers Formations. Tropical delta swamps & bayous dominated by conifers, ginkgos and tree ferns. Arid savannas dominated by ferns and low scrub with severe dry seasons. **Size:** WS >1.2m (4ft); BL 2.6m (8.5ft); TL unknown **Features:** Head high w/ tall, narrow snout. Wings highly reduced, but retaining large claws. Wings incapable of folding tightly, held forward or swept back against the sides. Legs short & powerfully built, tarsus very short. Large talons with large sickle claw. Stiffened tail probably used to maintain balance. **Biology:** Relatively slow-moving due to short tarsus, involved in the musculature for sickle claw. Preyed on contemporary ornithopods *Tenontosaurus tilletti,* mainly juveniles. May have mobbed prey in large flocks. Juveniles had longer wings & were possibly partially arboreal. Sickle-claw of juveniles more strongly curved than adults, & may have functioned in climbing.

Flat Bird-lizard Robber ***Saurornitholestes explanatus***
Time: 75 Ma ago **Location:** Alberta, Canada **Habitat:** Lower Dinosaur Park/Upper Oldman Formations. Seasonally arid lowland plains w/ braided river systems and small forests. **Size:** WS >75cm (2.5ft); BL 1.5m (5ft); TL unknown **Features:** Wings reduced but capable of grasping small objects w/ long, opposable digits in juveniles. Snout short. Juveniles stouter-skulled w/ longer limbs. **Biology:** Juvenile based on likely synonymous species *Bambiraptor feinbergorum*. Species *S. langstoni* (Sues 1978) likely synonymous with "Laelaps" *explanatus* (Cope 1876), which has priority (Mortimer 2010).

Stocky Dragon ***Balaur bondoc***
Time: 70 Ma **Location:** Transylvania, Romania **Habitat:** Sebes Formation, part of Hateg Island in the Tethys Sea. Warm, monsoonal with mountainous, dry forests and lakes in the uplands and swampy river deltas in the lowlands. **Size:** WS >80cm (2.6ft); BL ~1.5m (5ft); TL unknown **Features:** Wings large. Third finger lost, rest of wing relatively fused. Legs short & powerful, w/ sickle-claws on both second & third toes. **Biology:** The hallux, which was enlarged & raised parallel to the typical sickle claw, & the unusually short & stocky hind limbs, indicate a slow-moving ambush predator adapted to pin prey to the ground using the talons.

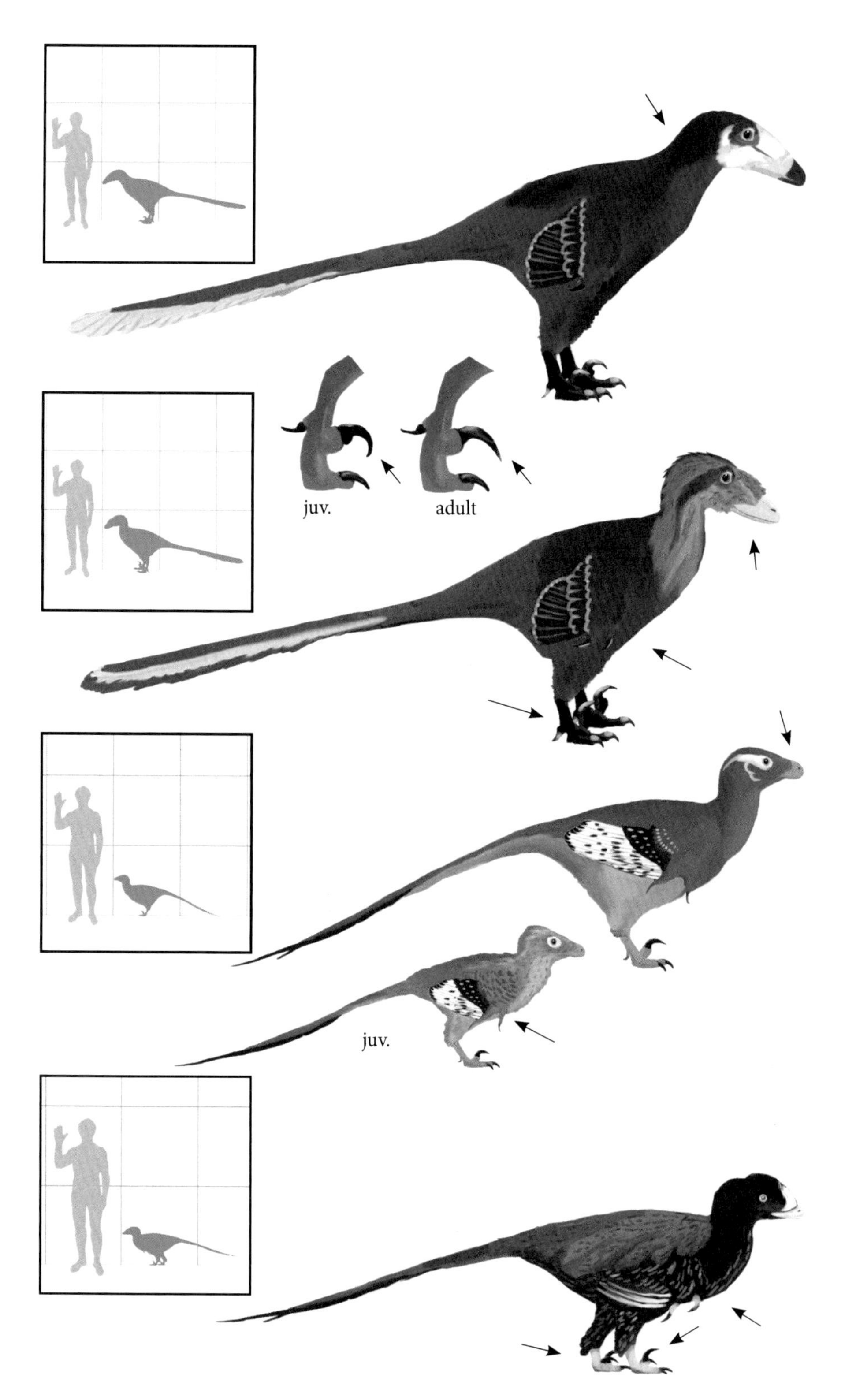
juv.
adult
juv.

Mongolian Swift Seizer ***Velociraptor mongoliensis***
Time: 75 Ma ago **Location:** Omnogovi, Mongolia & Inner Mongolia, China **Habitat:** Bayan Dzak & Tugriken Shireh, Djadochta Formation & Wulansuhai Formation. High desert. Dune fields and arid scrubland. **Size:** WS >80cm (2.6ft); BL 1.8m (6ft); TL unknown **Features:** Snout long and slightly concave. Neck moderately long. Wings short w/ large claws & long secondary feathers w/ strong shafts. Legs short w/ large sickle claw. Body deep, especially around the pelvis. Tail long & somewhat flexible. **Biology:** Hunted larger ornithischian prey including *Protoceratops andrewsi*, & also scavenged when possible. As in *Deinonychus*, juveniles had more strongly-curved sickle claws than adults. Unlike other eumaniraptorans, *Velociraptor* lacked a strong muscle connecting the shoulder to the upper wing, & so would not have been able to raise the wings above the horizontal (Parsons & Parsons 2009). In fact, they were probably not able to achieve even a fully horizontal position of the humerus. This lack of wing mobility is probably a secondary reversal. A specimen named *V. osmolskae* from Inner Mongolia differs only in pneumatic openings in the skull which are known to be variable in some other theropods & so is possibly a synonym.

Henan Luanchuan Thief ***Luanchuanraptor henanensis***
Time: Uncertain (Late Cretaceous) **Location:** Henan, China **Habitat:** Qiupa Formation **Size:** unknown; BL 2.7m (9ft); TL unknown **Features:** Head small w/ forward-facing eyes. Neck moderately long. Wings small but stout & powerfully muscled. Tail long, slender and somewhat flexible. **Biology:** The Qiupa Formation had a similar dinosaurian fauna to the Djadochta, including oviraptorid nesting sites. Probably similar in ecology to *Tsaagan*, though they were slightly more primitive.

White Monster *Tsaagan mangas*
Time: 75 Ma ago **Location:** Omnogovi, Mongolia & Inner Mongolia, China **Habitat:** Ukhaa Tolgod, Djadochta Formation. & Wulansuhai Formation. Dune fields and arid scrubland with nearby waterways. Nesting grounds for a wide variety of desert birds including caenagnathiformes, enantiornitheans and troodontids. **Size:** WS unknown; BL 1.2m (4ft); TL unknown **Features:** Snout taller & more box-shaped than *Velociraptor*, w/ larger teeth. Legs long compared to other species. Internal construction of the skull includes large maxillary fenestrae & foramina of the jugals. **Biology:** Probably preferred local prey such as oviraptorids and their nestlings. Contemporaries of similar *Velociraptor* species, these must have exploited different prey specializations or niches to avoid direct competition.

Mongolian Ada Lizard ***Adasaurus mongoliensis***
Time: 70 Ma ago **Location:** Bayankhongor, Mongolia **Habitat:** Nemegt Formation. Well-watered but arid near-desert environment dominated by low scrub, lakes, and dry woodland. **Size:** WS unknown; BL 2.7m (9ft); TL unknown **Features:** Head robust but snout unknown. Sickle claw very small. **Biology:** The unusually small sickle claws imply a reduced role for the hindlimbs/talons in prey capture.

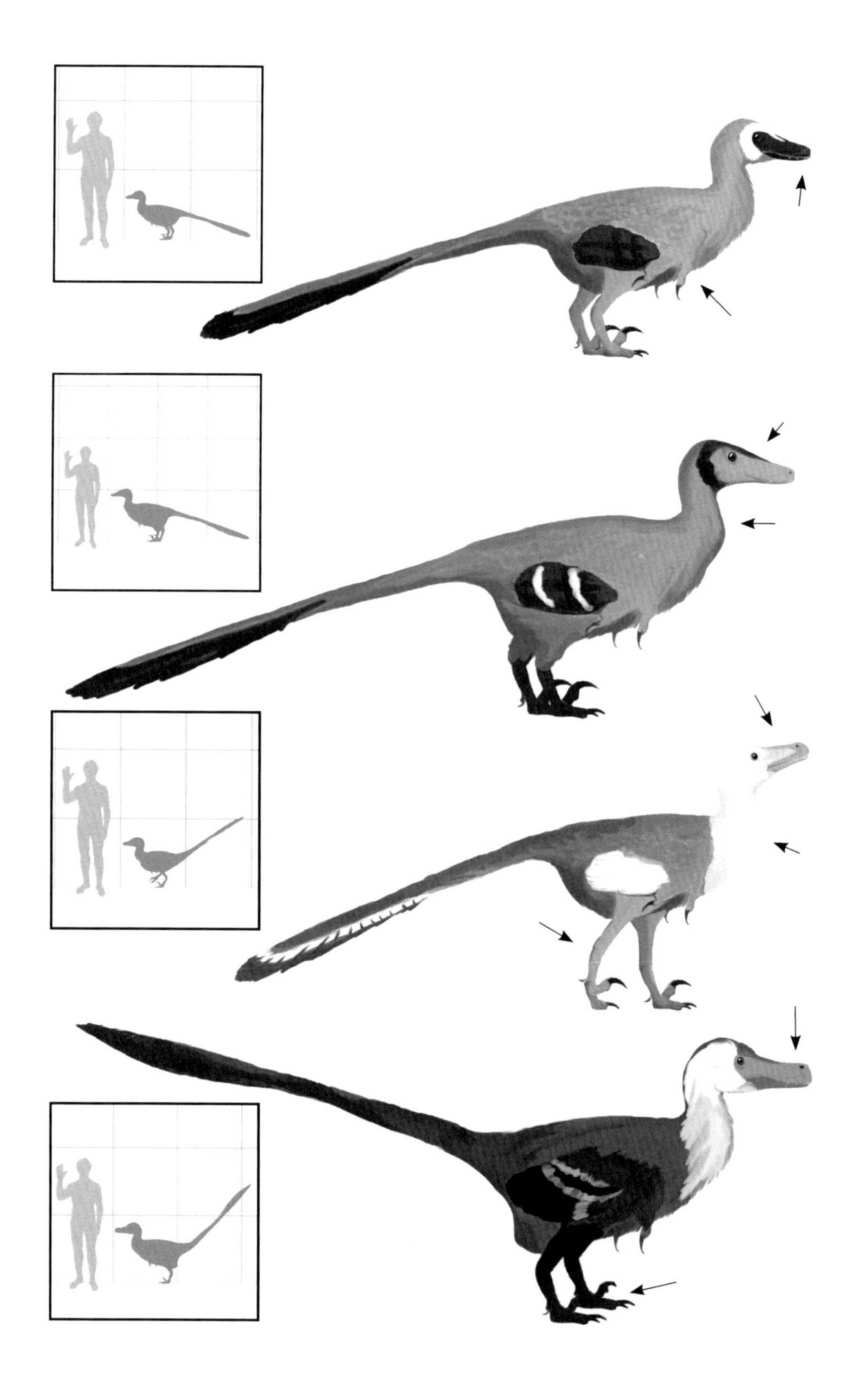

Unenlagiines

"Half birds", a highly specialized group of wading ornithodesmids adapted to heron-like ambush fishing. They are the only frond-tailed birds found below the equator, and the only definitive long-tailed birds from the southern continents. Because flight is only known to have definitively appeared in ornithothoraces (enantiornitheans & euornitheans), it is unknown how this group was able to disperse to South America and Madagascar. The possible primitive unenlagiine *Rahonavis ostromi*, though fragmentary, shows well-developed wings and may have had adequate enough powered flight to "island hop" between the northern and southern continents, subsequently spawning a secondarily flightless radiation of fishers. As in modern flightless birds, the feathers of the flightless unenlagiines probably became long and plumulaceous (downy or open-vaned).

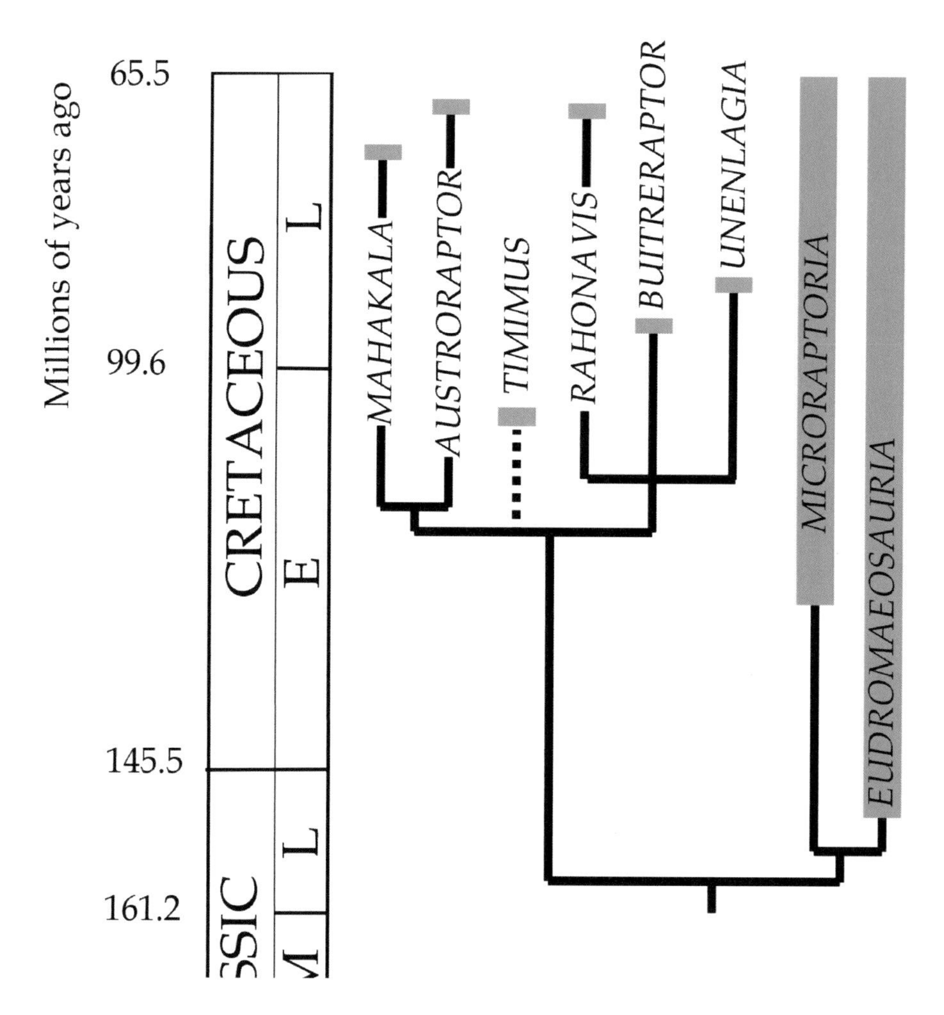

Above: Relationships of unenlagiines over time. Phylogeny approximated based on Senter et al. 2012.

Ostrom's Menace-from-the-Clouds Bird ***Rahonavis ostromi***
Time: 70 Ma ago **Location:** Mahajanga, Madagascar **Habitat:** Maevarano Formation **Size:** WS >80cm (2.6ft); BL ~80cm (2.6ft); TL unknown **Features:** Very small compared to other unenlagiines & relatively primitive, possibly representing a holdover from the ancestral form. Wings very large. Legs long, with unreversed hallux & hyperextendable sickle clawed second toe. Frond tail relatively short compared to the similar-looking microraptorians. **Biology:** The very long, powerful wings bearing quill knobs suggest these were capable of some limited flapping flight. Chiappe (2007) noted that like other unenlagiines, the wing could be raised more vertically than in related groups, allowing them to flap. A primitive powered flying stage for *Unenlagiinae* as a whole could help explain their distribution in the southern hemisphere, which was not connected by land to the north during the Cretaceous. *R. ostromi* was among the last of the aerial long-tailed birds.

Herman's Tim Mimic ***Timimus hermani***
Time: 106 Ma ago **Location:** Victoria, Australia **Habitat:** Eumeralla Formation, floodplains and braided rivers in a temperate and seasonally cold antarctic environment. **Size:** WS unknown; BL ~3m (9.8ft); TL unknown **Features:** Known only from isolated leg bones & vertebrae, they indicate a large animal w/ long, slender legs adept at running. **Biology:** Originally considered an ornithomimid, these are more likely eumaniraptorans, probably unenlagiines.

Comahue Half Bird ***Unenlagia comahuensis***
Time: 89 Ma ago **Location:** Neuquen, Argentina **Habitat:** Portezuelo Formation **Size:** WS unknown; BL ~3m (9.8ft); TL unknown **Features:** Very large flightless unenlagiines w/ small but highly mobile wings. Internally, large ridge on deltopectoral crest supporting strong wing musculature. **Biology:** Probably similar in ecology to other unenlagiines.

Paynemil's Half Bird ***Unenlagia paynemilli***
Time: 89 Ma ago **Location:** Neuquen, Argentina **Habitat:** Portezuelo Formation **Size:** WS unknown; BL ~2.2m (7.4ft); TL unknown **Features:** Similar to *U. comahuensis* & also known from fragmentary material. It differed in some internal anatomy of the shoulder girdle & hip bones, & appears to have been smaller. **Biology:** Probably similar in ecology to other unenlagiines.

Gonzalez's Vulture Roost Robber ***Buitreraptor gonzalezorum***
Time: 94 Ma ago **Location:** Neuquen, Argentina **Habitat:** Calenderos Formation **Size:** WS >70cm (2.3ft); BL 1.3m (4.4ft); TL unknown **Features:** Small, flightless unenlagiines w/ relatively large wings. Snout extremely long & narrow. wing claws short & digits nearly equal in length. Hand unusually short relative to very long humerus & radius/ulna. Relatively long legs. **Biology:** The very long, narrow snout suggests a piscivorous diet. Probably waded or stalked the shores of shallow waterways to ambush fish & small terrestrial vertebrates.

Cabaza's Southern Thief ***Austroraptor cabazai***
Time: 70 Ma ago **Location:** Rio Negro, Argentina **Habitat:** Allen Formation
Size: WS unknown; BL ~5m (16.4ft); TL unknown **Features:** Giant flightless unenlagiines. Snout long & narrow, w/ fluted teeth. Prominent ridge above the eyes. Wings very small & probably not readily visible among body feathers when folded. Legs long, w/ flat sickle claw on flattened second toe. **Biology:** Probably a generalist preferring fish, stalking river & lake shores like modern herons. Fluted teeth would have helped grasp slippery prey. Sickle claw may have helped tear larger prey or carcasses into manageable pieces. The extinction of the spinosaurids about 10 Ma prior may have allowed the unenlagiines, which were similar in ecology but much smaller, to have evolved larger sizes as they expanded into empty ecological niches.

Omnogov Mahakala ***Mahakala omnogovae***
Time: 75 Ma ago **Location:** Omnogov, Mongolia **Habitat:** Tugriken Shireh, Djadochta Formation, highly arid Gobi Desert scrub and dune fields. **Size:** WS >20cm (7.8in); BL 65cm (2.1ft); TL unknown **Features:** Tiny, ground-dwelling deinonychosaurians. Wings extremely reduced & possibly not visible externally. Legs slender. Tail shallow but very wide & flattened in appearance. Internally characterized by a flattened, broad ulna & large femoral crest. **Biology:** This species preserves some features of the earliest deinonychosaurians, such as very small size, but others, such as the extremely small wings, may represent reversals. Probably foraged for small prey among scrub or buried under the sand. As a very late surviving, terrestrial member of its lineage (possibly an unenlagiine) with a foraging lifestyle, the plumage is restored similarly to modern small flightless birds like kiwi.

Troodontids

Troodontids, "wounding teeth", are a relatively small group of deinonychosaurians, but are in some ways more similar to modern birds. They tend to be more slender in build, with longer legs well suited to running. They usually exhibit long, narrow snouts filled with small, leaf-shaped teeth, and at least some may have been omnivorous. They are generally smaller-winged and shorter-tailed than other large frond-tailed birds.

Advanced troodontids, the troodontines, were all similar to each other in appearance and size. The general troodontine body plan was very successful in the Late Cretaceous, with similar species dispersed through Asia and North America. Many more species are known only from teeth, such as *Pectinodon bakkeri*, which shows that troodontines survived until the very end of the Cretaceous period.

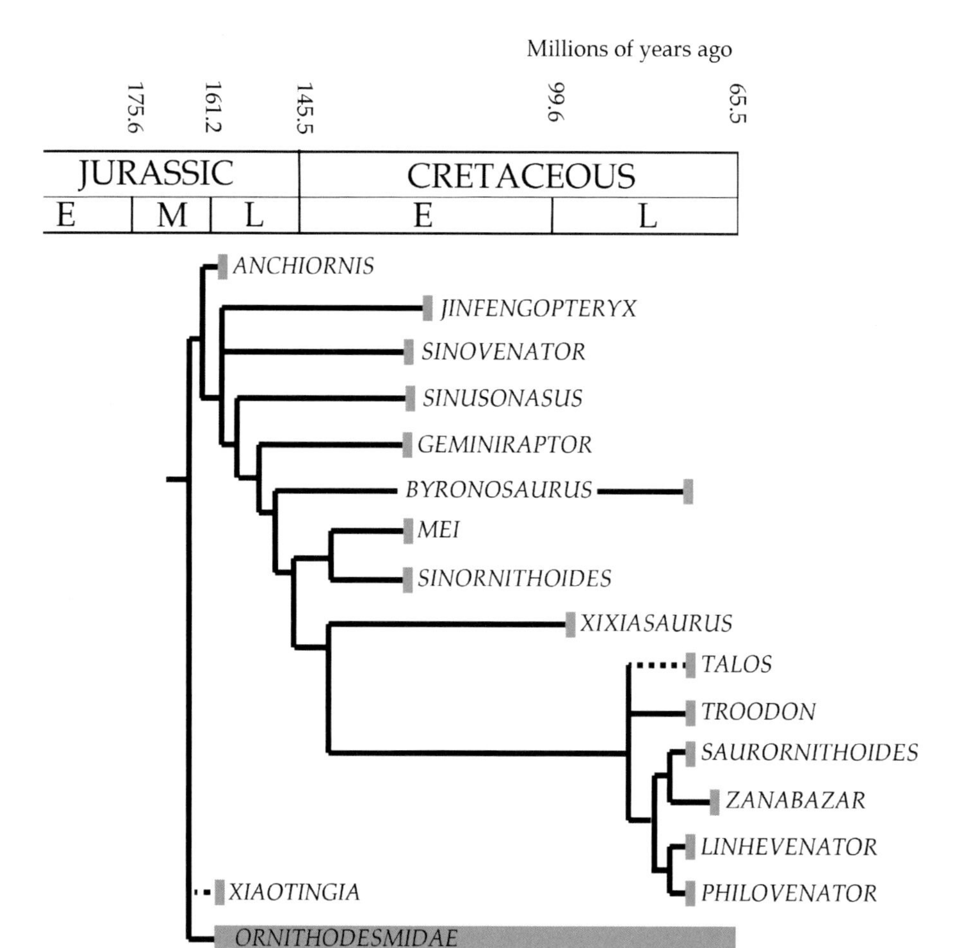

Above: Relationships of troodontids over time. Phylogeny approximated based on Senter & al. 2012.

Elegant Golden Phoenix Feather ***Jinfengopteryx elegans***
Time: 122 Ma ago **Location:** Hebei, China **Habitat:** Qiaotou Member, Huajiying Formation **Size:** WS >35cm (1.1ft); BL 55cm (1.8ft); TL ~65cm (2.1ft) **Features:** Head triangualar but rounded in profile. Wings small w/ large claws. Torso short. Legs long & lacking feathers. Tail long, with feather frond extending to base. Rectrices up to ~10cm long at the tail tip, much shorter near base. Internally, third wing digit reduced & fused. **Biology:** Omnivorous or herbivorous, foraged on the ground for seeds, nuts, & possibly insects & small vertebrates.

Henan Xixia Lizard ***Xixiasaurus henanensis***
Time: Uncertain (Late Cretaceous) **Location:** Henan, China **Habitat:** Majiacun Formation **Size:** WS unknown; BL ~1.2m (3.9ft); TL unknown **Features:** Snout long & rounded with U-shaped jaws (as seen from below), slightly indented on the sides. **Biology:** The solidly-constructed, box-shaped skull (as well as the large overall body size) may have evolved to reduce stresses when feeding on larger prey items.

Soundly Sleeping Dragon ***Mei long***
Time: 124.5 Ma ago **Location:** Liaoning, China **Habitat:** Lower Yixian Formation, temperate conifer and ginkgo forest set among a series of lakes fed by streams and runoff from a nearby range of active volcanic mountains. **Size:** WS >30cm (11.8in); BL 45cm (1.5ft); TL unknown **Features:** Head large & rounded w/ short, flattened snout. Teeth numerous, small, unserrated, and closely-packed, filling much of the upper jaw extending to below the eye. Wings relatively short. Legs very long, w/ long tarsus & relatively short femur. Tail long & flexible. Internally, wishbone U-shaped as in oviraptorids. **Biology:** Known from specimens found in roosting postures, probably sheltered against the volcanic ash which buried them. Generally similar to the contemporary *Sinovenator*. Despite juvenile-like appearance including large head/eyes & long legs, studies of bone histology show that these specimens were mature (Gao & al. 2012).

Chang's China Hunter ***Sinovenator changi***
Time: 124.5 Ma ago **Location:** Liaoning, China **Habitat:** Lower Yixian Formation (see above) **Size:** WS unknown; BL 1.1m (3.6ft); TL unknown **Features:** Head triangular. Legs long, wings small. Tail long & flexible, in some roosting specimens wrapped around the body. **Biology**: Probably hunted small game in & around Yixian lake systems, but most fossils have been found in non-lake ash deposits, suggesting a primarily inland, terrestrial habitat. The similar *Sinusonasus magnodens* is probably a synonym (Turner & al. 2012).

Suarezes' Twin Seizer ***Geminiraptor suarezorum***
Time: 126 Ma ago **Location:** Utah, USA **Habitat:** Yellow Cat Member, Cedar Mountain Formation. Open, marshy mud flats. **Size:** WS unknown; BL ~1.3m (4.4ft); TL unknown **Features:** Mid-sized troodontids known from a single snout bone (maxilla). High, rounded snout characterized internally by extensive air-filled cavities and prominent, elongated openings. Tooth sockets square & separated by small walls of bone. **Biology:** The broad snout & odd tooth arrangement may indicate an unusual method of feeding among deinonychosaurians.

Young's Chinese Saurornithoid ***Sinornithoides youngi***
Time: 125 Ma ago **Location:** Inner Mongolia, China **Habitat:** Ejinhoro Formation **Size:** WS >50cm (1.6ft); BL 1.3m (4.4ft); TL unknown **Features:** Head small w/ somewhat pointed snout. Neck relatively long. Body long with very small wings & small wing claws. Legs very long with relatively small talons. Tail relatively short. **Biology:** The extremely long legs & very small wings indicate an exclusively terrestrial habitat. Likely fast runners, the long legs and slender, pointed snout may suggest that these were wading, aquatic foragers.

Byron Jaffe's Lizard ***Byronosaurus jaffei***
Time: 75 Ma ago **Location:** Omnogovi, Mongolia **Habitat:** Ukhaa Tolgod, Djadochta Formation. Dune fields and arid scrubland with nearby waterways. Nesting grounds for a wide variety of desert birds. **Size:** WS unknown; BL ~1.4m (4.6ft); TL unknown **Features:** Snout long & narrow. Nestlings very small & dissimilar in appearance, having triangular, pointed heads. Nested nearby to caenagnathiformes (*Citipati osmolskae*) and enantiornitheans (*Gobipteryx minuta*) among sand dune fields. **Biology:** Possibly hunted small vertebrates such as lizards. Asymmetrical ear openings similar to owls allowed them to pinpoint small prey hidden in brush or buried in sand.

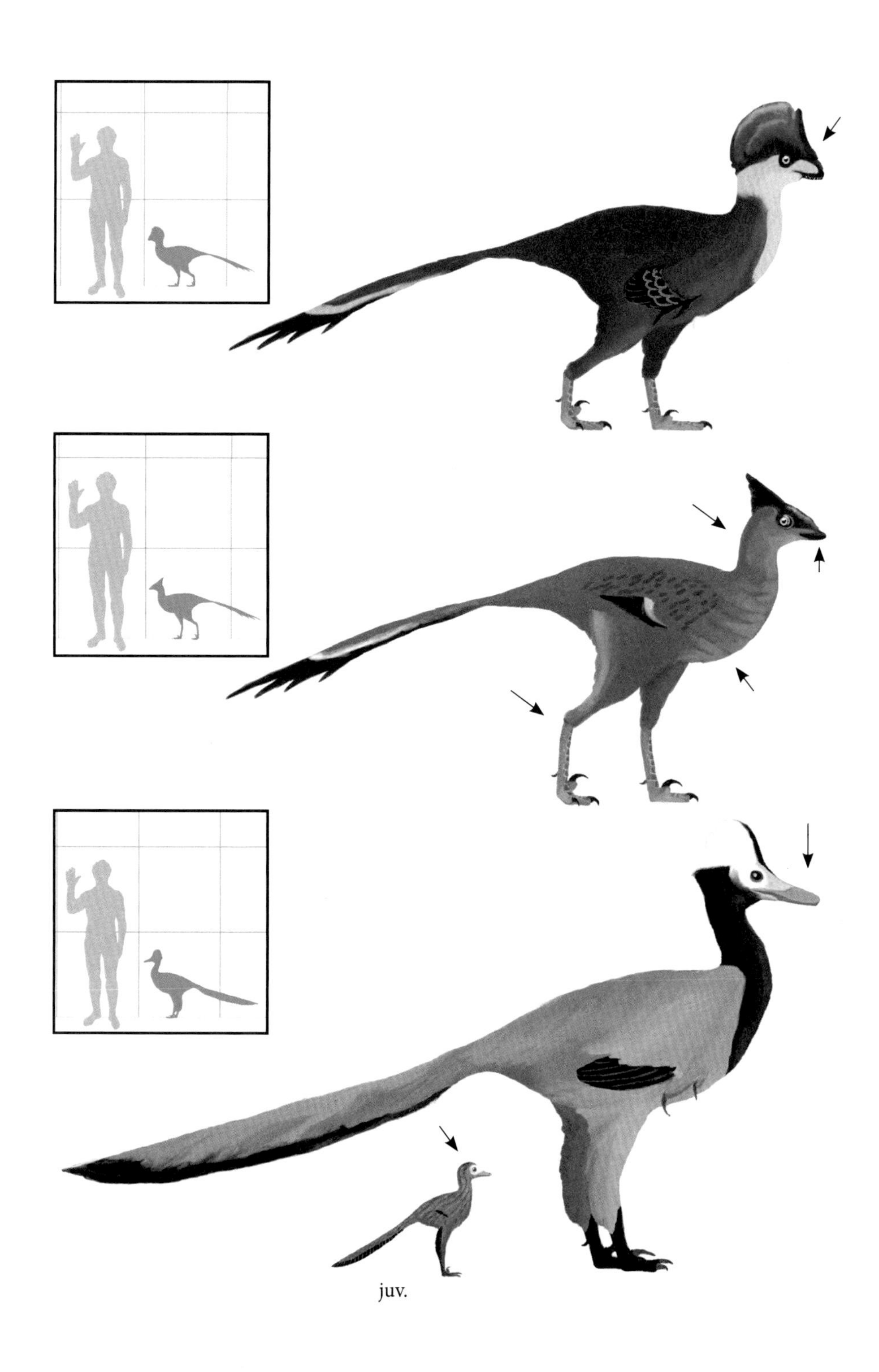
juv.

Mongolian Bird-like Saurian *Saurornithoides mongoliensis*
Time: 75 Ma ago **Location:** Omnogovi, Mongolia **Habitat:** Bayan Dzak, Djadochta Formation. Arid desert scrubland. **Size:** WS unknown; BL 1.5m (4.9ft); TL unknown **Features:** Mid-sized relative to contemporary species. Snout long & narrow, but broader than adult *Byronosaurus*. Teeth w/ large, coarse serrations. Legs long & slender. **Biology:** Long tarsi indacte that these were fast runners. Probably fed on small vertebrates such as lizards & hatchling dinosaurians foraged from the dunes.

Young Zanabazar *Zanabazar junior*
Time: 70 Ma ago **Location:** Omnogovi, Mongolia **Habitat:** Nemegt Formation (see above) **Size:** WS unknown; BL 2.5m (8.2ft); TL unknown **Features:** Large troodontids. Snout long and narrow. Large number of teeth (20 in upper jaw, 35 in lower jaw) w/ large, coarse serrations. Legs long & slender. **Biology:** Possibly omnivorous generalists, preferring to take small prey but supplementing with plant material, invertebrates, etc.

Tan's Linhe Hunter *Linhevenator tani*
Time: 75 Ma ago **Location:** Inner Mongolia, China **Habitat:** Wulansuhai Formation **Size:** WS unknown; BL ~1.7m (5.6ft); TL unknown **Features:** Snout relatively long but w/ broad skull caused by flared jugal bones. Wings small but robust & powerful. Sickle claw unusually large among troodontids. **Biology:** Thr large sickle claw likely evolved in parallel w/ eudromaeosaurians, indicating a similar prey capture strategy in this species. The arms are short as in other troodontids but especially powerful, possibly used in digging, climbing, or clinging to large prey.

Phil Currie's Hunter *Philovenator curriei*
Time: 75 Ma ago **Location:** Omnogovi, Mongolia & Inner Mongolia, China **Habitat:** Djadochta & Wulansuhai Formations. Arid desert scrubland. **Size:** (juvenile) WS unknown; BL ~75cm (2.5ft); TL unknown **Features:** Very small troodontines. Head triangular w/ narrow, pointed snout. Head large, legs long & slender, and, unusually, wider from front to back than from side to side. Sickle claw unusually small. **Biology:** Though known only from subadult (~two-year-old) specimens, some of which were initially considered juvenile *Saurornithoides mongoliensis*, these are distinct in several anatomical characteristics, including a large sheet-like process on the tibia, presumably for anchoring powerful leg muscles. They appear to be more closely related to *Linhevenator*. Additional specimens from the same time & general area, known as the "Zos Canyon Troodontid", are probably the same species (Mortimer 2010). Small sickle claw may indicate less reliance on the foot talons in prey capture.

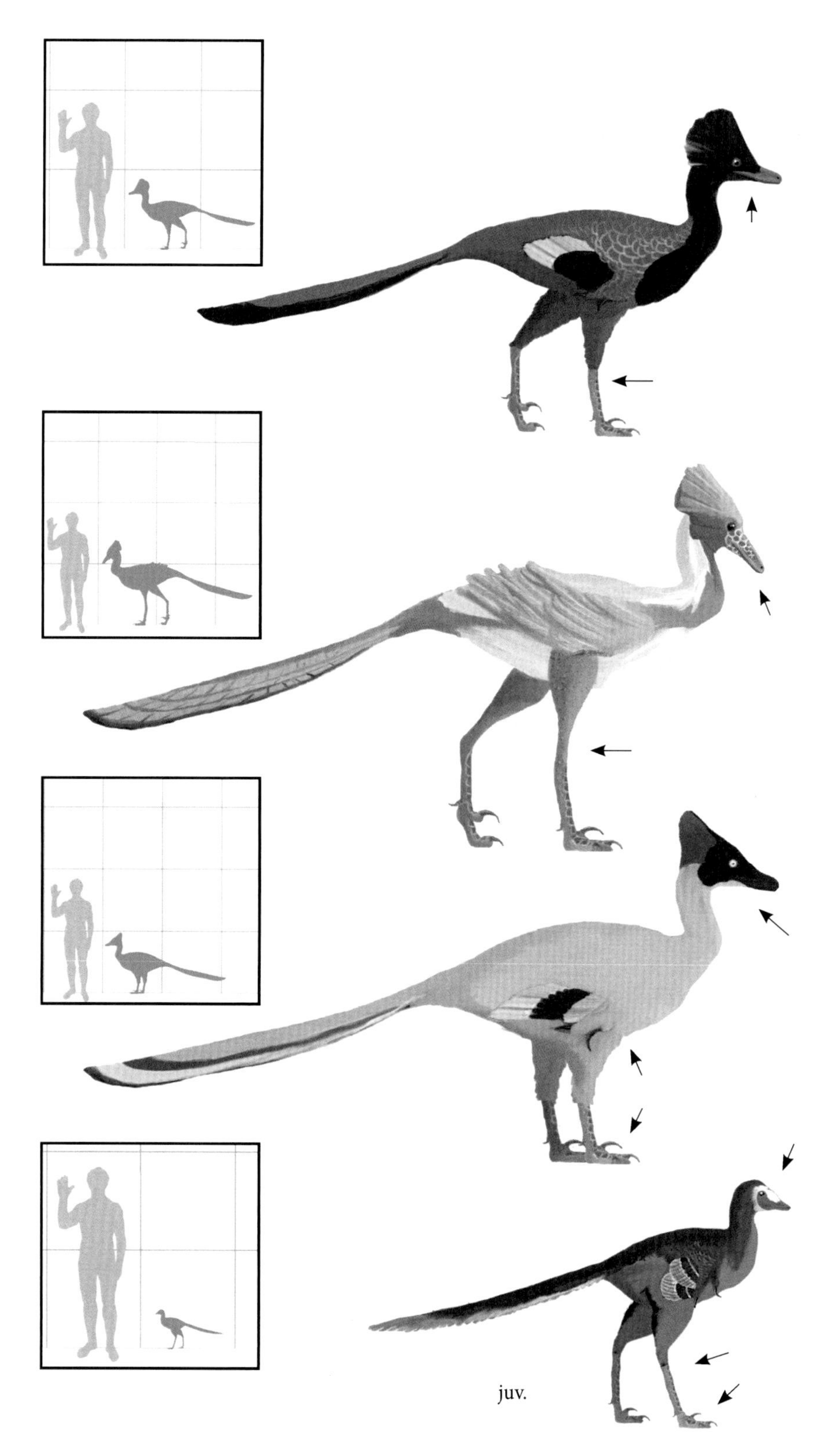
juv.

Sampson's Talos *Talos sampsoni*
Time: 76 Ma ago **Location:** Utah, USA **Habitat:** Kaiparowits Formation. Floodplain dominated by large, seasonally flooding rivers. Occupied wide muddy and sandy riverbanks. **Size:** WS unknown; BL ~1.6m (5.2ft); TL unknown **Features:** Mid-sized to large troodontines. Wings very short w/ slender forearms. Legs slender but shorter than contemporary Asian species. Sickle claw relatively small. **Biology:** Like other troodontids, these birds may have used their large recurved talons to pin prey, likely using a single foot. An injury to the second toe in the first known fossil specimen supports the idea that the toe was used routinely to pin down small, struggling animals.

Handsome Wounding Tooth *Troodon formosus*
Time: 75 Ma ago **Location:** Montana, USA & Alberta, Canada **Habitat:** Judith River, Dinosaur Park and Horseshoe Canyon Formations. Seasonally arid lowland plains dominated by braided river systems and small forests. **Size:** WS unknown; BL ~2.3m (7.5ft); TL unknown **Features:** Large troodontines. Head long w/ long but broad snout. Wings very small. Legs moderately long but stouter than contemporary Asian species. Sickle claw large & flattened similar to eudromaeosaurians, probably for use in predation. **Biology:** Carnivorous, but possibly supplementing with fruit or seeds. Small teeth classified in the species *Paronychodon lacustris* possibly represent juvenile *T. formosus*. While fossils attributed *T. formosus* have been found in a huge range of formations (including very large specimens from Alaska) and spanning many millions of years (up to 66 million years ago), most of these are based on extremely fragmentary remains, and almost certainly petrain to different species.

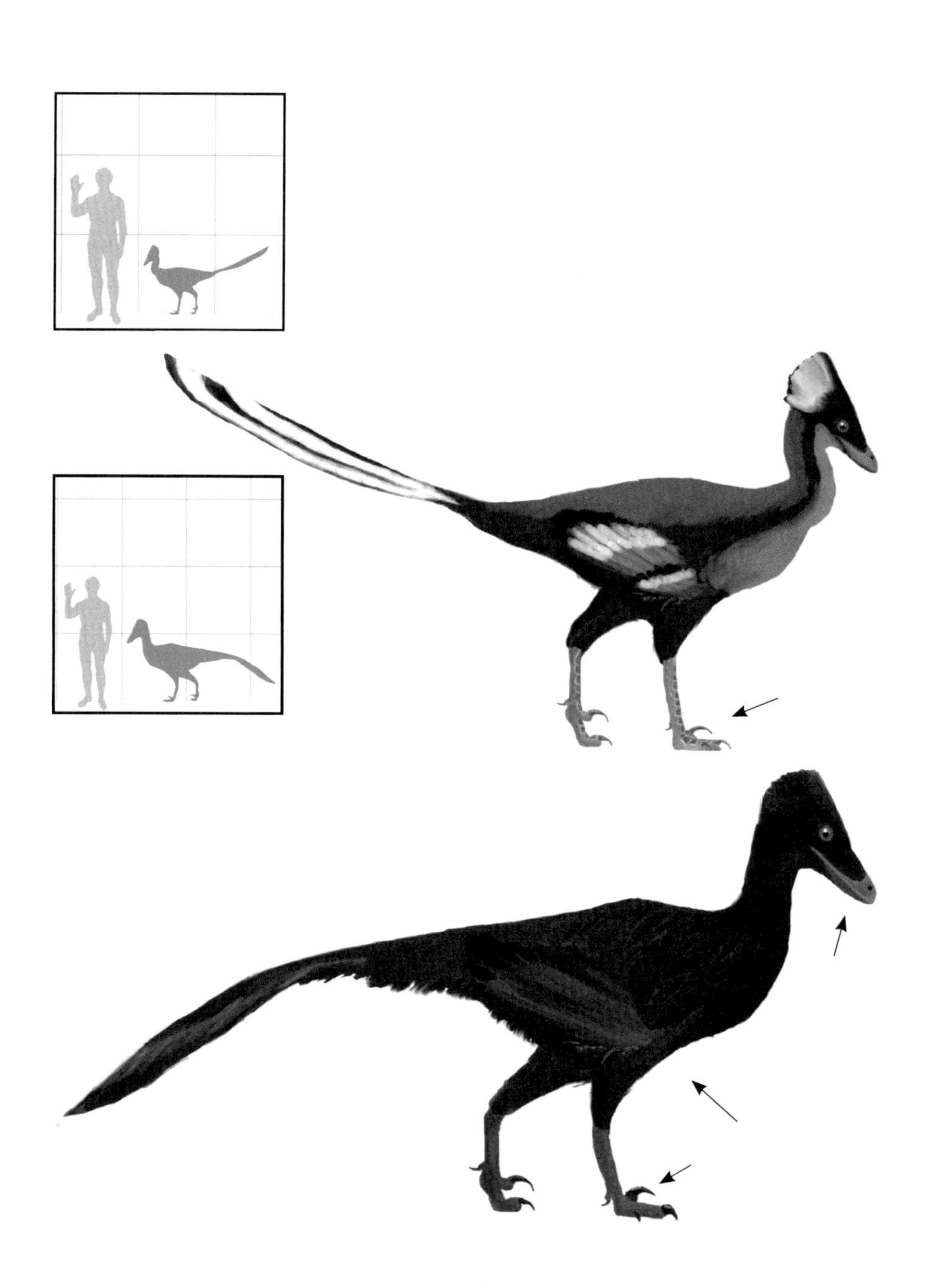

Basal Avialans

The lineage leading to modern birds probably diverged from the deinonychosaurian lineage during the mid- or early Jurassic. All birds closer to *Aves* than to deinonychosaurians are called avialans ("winged birds"), though that name has also been used for a group based on a physical character (the presence of wings used for flight) and so might be replaced by the PhyloCode.

Successive groups of avialans acquired more and more characteristics of modern birds, including the fusing together of the bones in the wing, lengthening the forearm relative to the hind limbs, and shortening of the tail. Only a few of these more advanced long-tailed birds are known, including the Chinese forms *Jeholornis* and *Yandangornis*. The most advanced known bird fossil that can reasonably be considered "long-tailed" is the one named "*Zhongornis haoae*", which has a very shortened tail, though not as short as *Confuciousornis*, and lacking a pygostyle or any caudal fusion. Interestingly, it is likely that the "*Zhongornis*" specimen is simply a juvenile confuciousornithid, which would have implications for the evolution of short tails: a short-tailed bird that retained a long tail as a chick, almost as tadpoles reduce their tails as they mature into frogs.

The avebrevicaudans, "short-tailed birds", represent the first birds to shorten their tails to the point that they possessed ten or fewer vertebrae. The last few vertebrae were usually fused together into a single solid structure, similar to the pygostyles of true birds (euornitheans). However, there is no evidence that these tails possessed mobile feather fans--on the contrary, most known primitive avebrevicaudans appear to have had only a few pairs of streamer-like rectrices, if any.

Confuciusornithids and other primitive short-tailed birds generally had long, large wings, though they lacked well-developed breast muscles and could not lift their wings very far above their backs, preventing strong flapping or ground-based takeoffs. The low, inward and forward-facing halluces and large, strong wing claws imply that they instead climbed tree trunks to reach gliding or flying height. Many species had toothless beaks, which evolved independently of those found in modern birds.

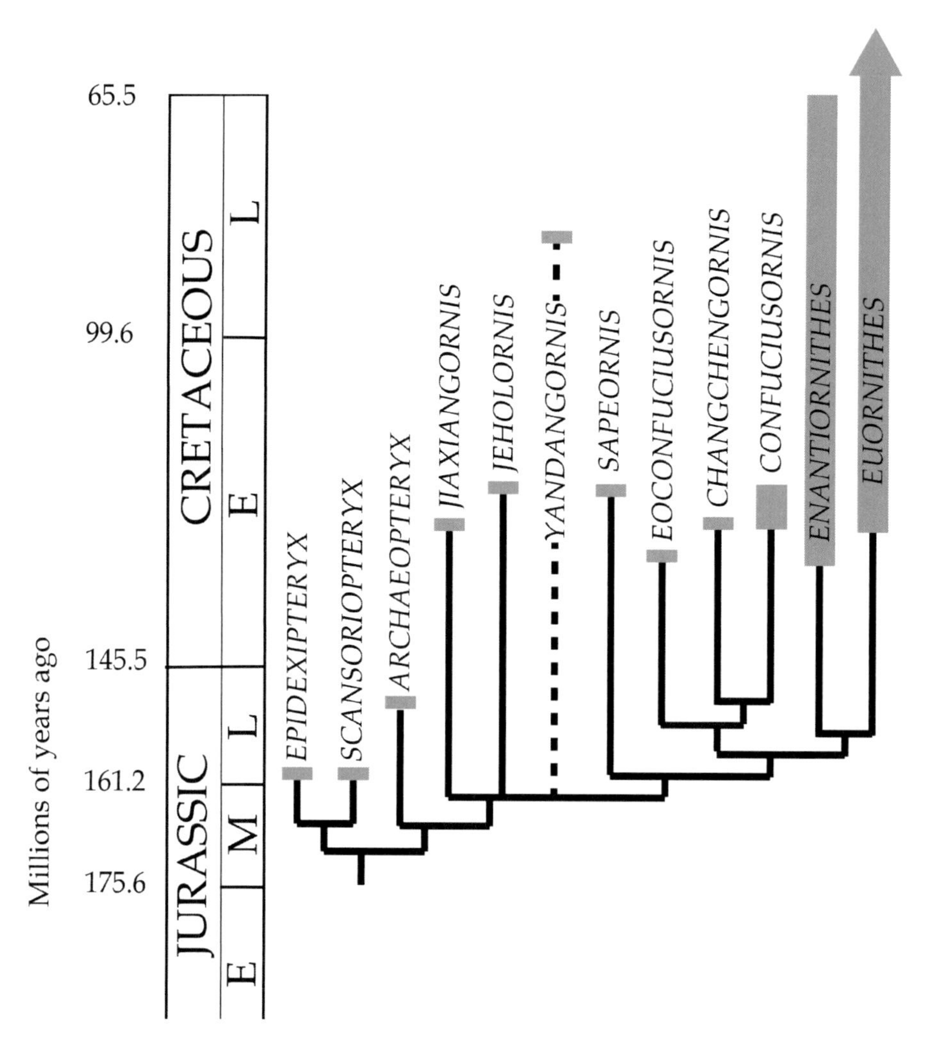

Above: Relationships of avialans over time. Phylogeny approximated based on O'Connor & Zhou 2012.

Heilmann's Climbing Wing ***Scansoriopteryx heilmanni***

Time: 155 Ma ago **Location:** Inner Mongolia & Liaoning, China **Habitat:** Daohugou Formation **Size:** (juvenile) WS >10cm (3.9in); BL 21cm (8.3in); TL 22cm (8.6in) **Features:** Head round w/ short, rounded snout. Teeth forward-pointing. Wings large w/ extremely long third finger. Legs short w/ four forward-pointing toes. Tail long w/ frond-like feathers restricted to the tip of the tail. **Biology:** Known only from hatchling specimens. Had an unusual suite of primitive & advanced features that initially led some scientists to conclude that not only were these the ancestors of all birds, but were more primitive than most other dinosaurs (mainly due to a hip lacking the characteristic perforated socket of all dinosaurs, which may be explained by the juvenile nature of the specimens). Unlike almost all other dinosaurs, the minor digit was longer than the major. Extending the third finger & its claw past the primary feathers would have allowed greater use in climbing or probing bark. Other features may also have been related to a climbing lifestyle. The hallux was unreversed but level w/ other toes & forward-facing, and may have aided in climbing tree trunks. These animals were probably insectivorous & more highly arboreal than their relatives. Likely able to glide or parachute, remiges in the juvenile specimens display a distinctive herringbone pattern indicative of vanes w/ barbules. The remiges appear to be attached to the third digit in the fossil, suggesting that the second & third digits were fused together with soft tissue in life.

Hu's Display Feather ***Epidexipteryx hui***

Time: 155 Ma ago **Location:** Inner Mongolia, China **Habitat:** Daohugou Formation (see above) **Size:** WS n/a; BL 30cm (11.8in); TL >45cm (1.5ft) **Features:** Large, forward-pointing teeth. Head tall & box-like. May have had a long third finger similar to *Scansoriopteryx*. Tail short, bearing four long ribbon-like feathers. Because they lacked a true pygostyle, the tail feathers were probably largely immobile relative to the tail & permanently fanned out. **Biology:** Wings highly degenerate & lacking primaries, indicating loss of gliding ability. Probably insectivorous. The short tail of *E. hui* was fused, a condition which probably evolved independently of other birds and is related to tail shortening. This was used as support for a dramatic set of four highly elongated tail feathers (ETFs). Unlike modern feathers but similar to confuciusornithids & enantiornitheans, the central quill of these ETF was broad, flat, and lacked a vane of barbs, forming a single ribbon-like sheet.

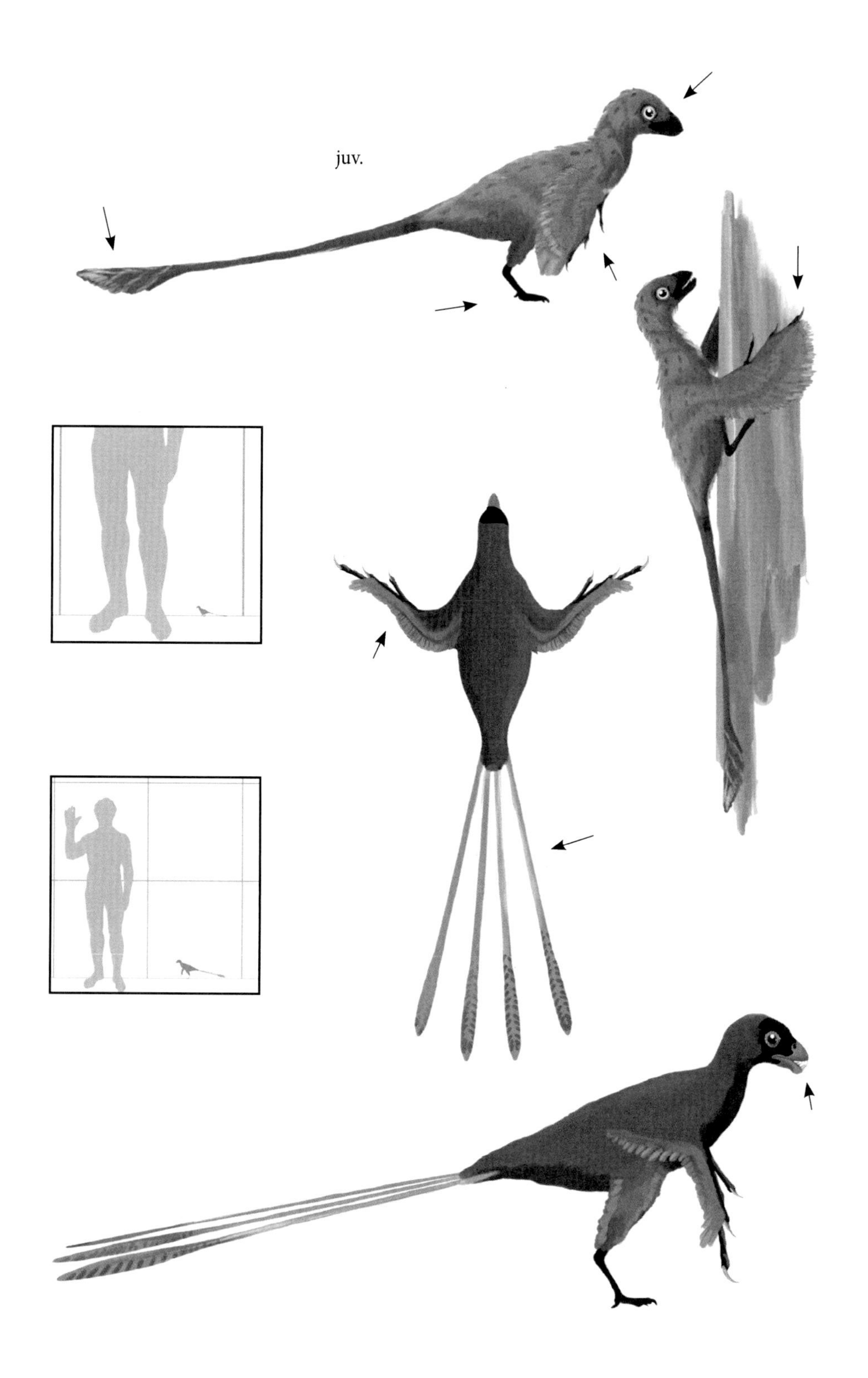
juv.

Eastern Jixiang Bird ***Jixiangornis orientalis***
Time: 124.5 Ma ago **Location:** Liaoning, China **Habitat:** Lower Yixian Formation **Size:** WS ~1m (3.3ft); BL 70cm (2.3ft); TL unknown **Features:** Head triangular. Teeth either absent or very small & not preserved in known specimens. Tip of lower jaw expanded & prominent. Body long. Wings long but details of feathers unknown. Hallux not reversed, but may have pointed medially. Tail moderately long w/ rectrices restricted to the tip, shorter than that of the similar *Jeholornis prima*. Details of rectrices unknown, but may have been palm-like similar to *J. prima*. **Biology:** Very similar to *J. prima*, differed in shorter tail & apparent lack of teeth.

First Jehol Bird ***Jeholornis prima***
Time: 120 Ma ago **Location:** Liaoning, China **Habitat:** Jiufotang Formation. **Size:** WS 1m (3.3ft); BL 60cm (2ft); TL 70cm (2.3ft) **Features:** Head triangular, w/ few small teeth in upper & lower jaws. "Chin" prominent, pointed in some specimens, but rounded in others. Wings long & somewhat pointed. Hallux not reversed. Tail long w/ short rectrical frond at the tip. Recrices narrow & curved outward w/ pointed non-overlapping tips. **Biology:** Known to have eaten seeds & probably foraged mainly on the ground. Small rectrical fronds lacked aerodynamic features, probably for display. *Jeholornis palmapenis* was named based on contemporary speciemens, distinguished among other things by presence of small teeth. However, this is probably preservational & the two likely represent the same species. The name *Jeholornis prima* was published on July 25, 2002 in a weekly journal, while another likely synonym, *Shenzhouraptor sinensis,* was published in a monthly journal with no day date. The ICZN rules that its publication date is therefore to be considered July 31, 2002 barring an actual print date. While a press release accompanying the *Shenzhouraptor* description was dated July 23, 2002, there is no published evidence that it & the print journal appeared at exactly the same time, so *Jeholornis* must be considered the earlier name.

Long-tailed Yandang Bird ***Yandangornis longicaudus***
Time: 85 Ma ago **Location:** Zhejiang, China **Habitat:** Tangshang Group **Size:** WS >60cm (2ft); BL 60cm (2ft); TL ~65cm (2.1ft) **Features:** Head long, w/ a long and pointed but robust toothless bill. Wings long. Legs very long w/ strong muscle/ligament attachment at the ankles, likely good runners. Foot claws small & slightly curved. Hallux small, high on tarsus & unreversed. Tail thin & short. Rectrices restricted to tip. **Biology:** Probably inhabited muddy river or lake shores feeding on fish & invertebrates.

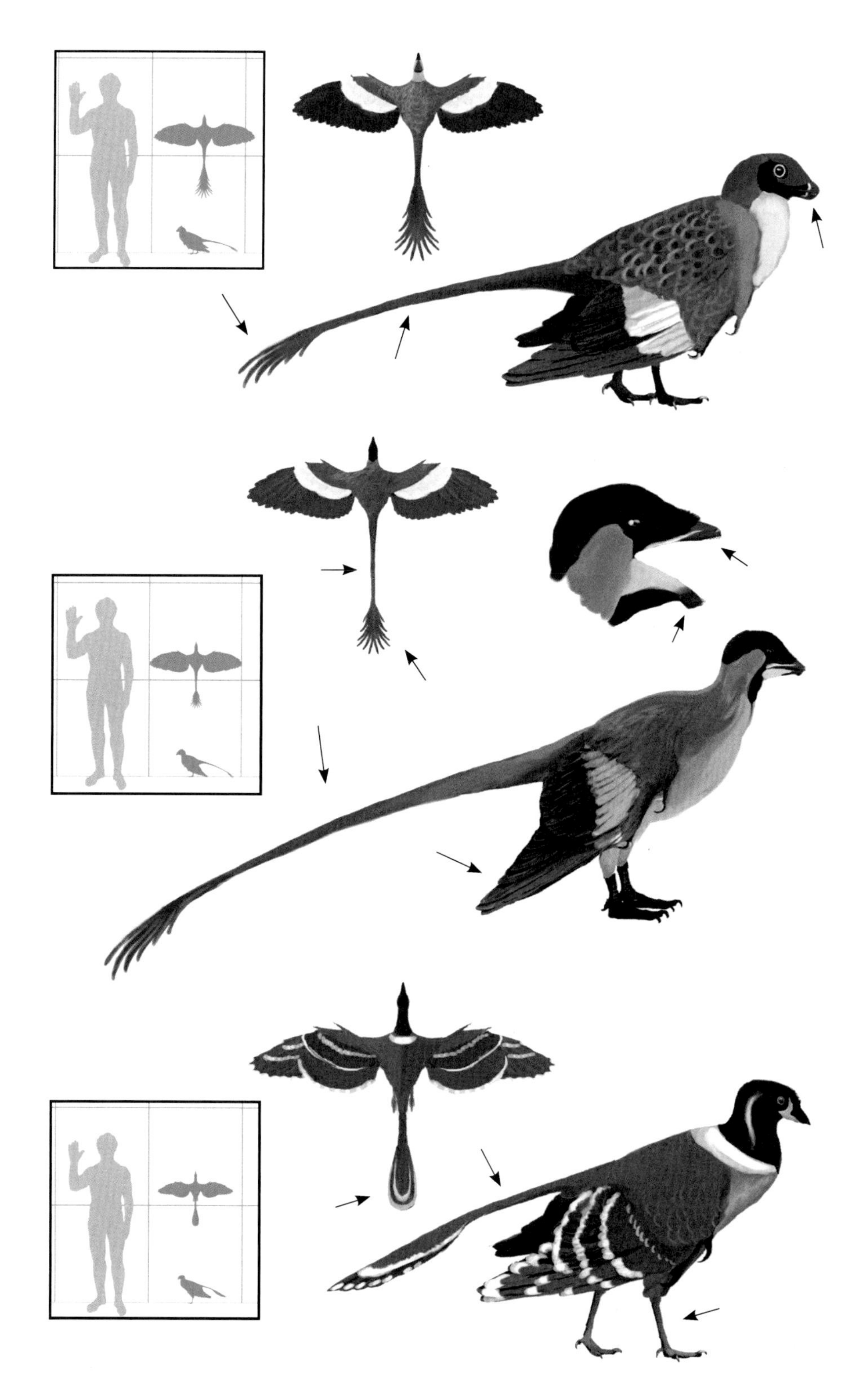

Holy Confucius Bird ***Confuciusornis sanctus***
Time: 124.6-122 Ma ago **Location:** Liaoning, China **Habitat:** Yixian Formation **Size:** WS 63cm (2ft); BL 25cm (10in); TL 50cm (1.6ft) **Features:** Head round w/ long, sharply pointed bill. Body black to dark grey, mottled w/ chestnut or reddish brown. Wings extremely long & pointed in adults, slightly shorter in immature specimens. Wing claws very large, strongly curved & unfeathered. Primaries white, secondaries & coverts grey to black. Legs very short w/ small feet and medial hallux. Males ribbon-tailed w/ single pair of very long rectrices. Females lack rectrices, tail consisting of tiny coverts (also present in males). **Biology:** Young chicks have longer, bony tails which fuse & shorten as they mature. Males develop rectrices at a young age at or near hatching; rectrices are proportionally longer in smaller individuals. Growth slow during 1st year or two, & rapid from mid- to full-size. Wing anatomy radically different from avians; breast small & undeveloped, breastbone bears only a small, cartilaginous keel. Large opening in humerus indicates unique musculature allowing some degree of flapping, but angle at which wings could be raised above back limited by shoulder anatomy. Probably scaled trees using large wing claws, may have been able to navigate canopy using medial hallux. May have glided between trees using weak flapping, perhaps swooping in flocks over lakes where many have been preserved. By far the most common bird species in the Yixian, they were possibly colonial, forming huge flocks, some of which died en masse during volcanic eruptions & were buried simultaneously on the lake bottoms. Young juveniles absent from large fossil assemblages, indicating that birds below mid-size lived in isolation or in more inland environments before joining flocks. Diet is unknown in this species.

Du's Confucius Bird ***Confuciusornis dui***
Time: 124.6 Ma ago **Location:** Liaoning, China **Habitat:** Yixian Formation **Size:** WS 52cm (1.7ft); BL 17cm (7in); TL 40cm (1.3ft) **Features:** Very similar in overall anatomy to *C. sanctus*. Differed in smaller adult size, an upturned bill tip, & a much smaller alular claw. **Biology:** The differing bill shape is a strong indicator of a unique diet, though no stomach contents are known. Later confuciusornithids are known to have fed on small fish, & this seems like a good possibility for the bill function of *C. dui*.

Zheng's Dawn Confucius Bird ***Eoconfuciusornis zhengi***
Time: 131 Ma ago **Location:** Hebei, China **Habitat:** Sichakou Member, Huajiying Formation **Size:** WS 38cm (1.2ft); BL 17cm (7in); TL 40cm (1.3ft) **Features:** Beak short & pointed, head triangular. Wings pointed but broad & relatively stout, w/ long secondaries. Legs short with partially reversed hallux. Ribbon-tail consisting of a single pair of very long rectrices. **Biology:** Dark preservation of melanin in the only known specimen suggests a very dark, possibly black uniform coloration. Similar in most respects to the *Confuciusornis*, but much earlier chronologiacally. Also differed in small shoulder bones (coracoids) & relatively solid vertebrae.

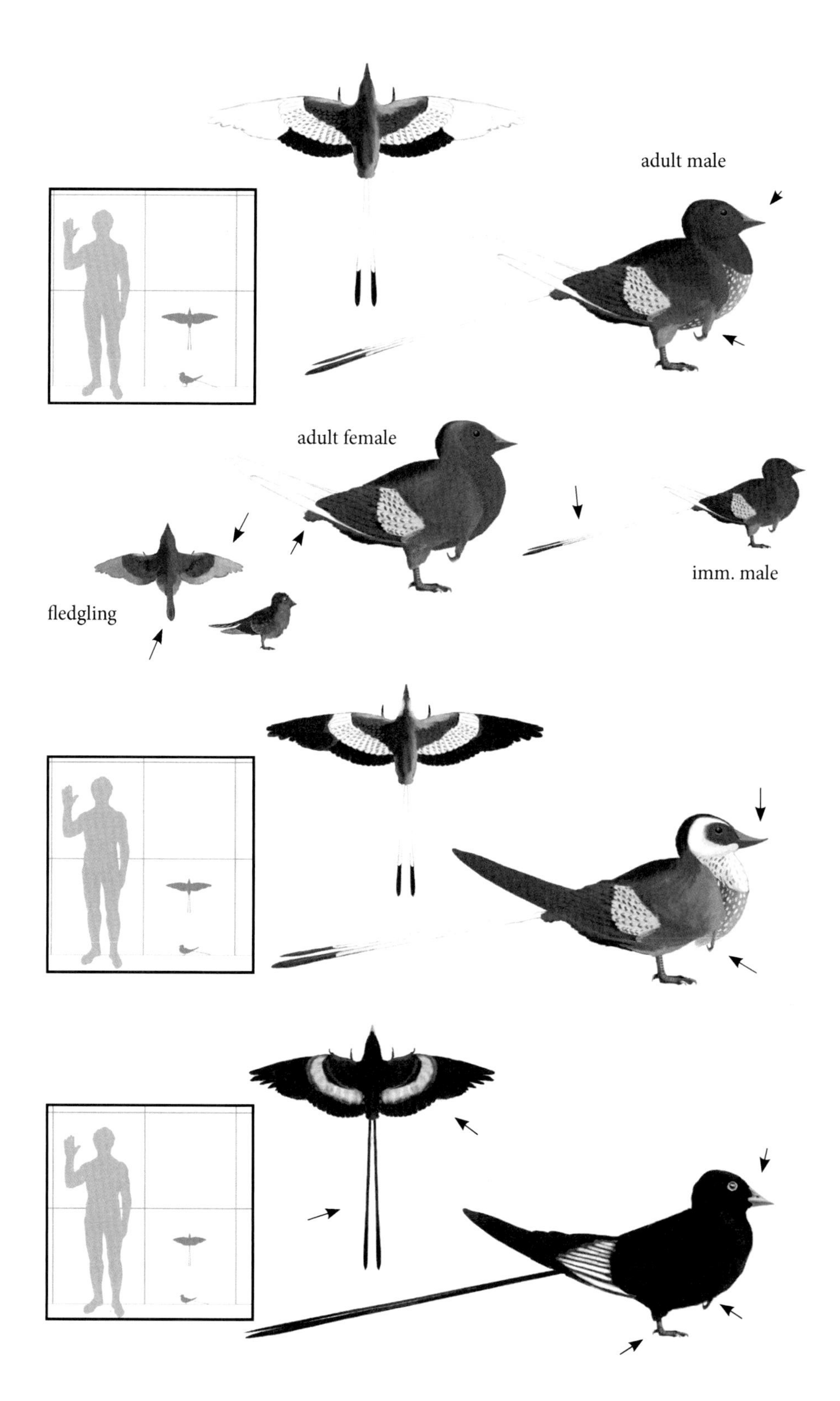
adult male
adult female
imm. male
fledgling

Jianchang Confucius Bird *Confuciusornis jianchangensis*
Time: 120 Ma ago **Location:** Liaoning, China **Habitat:** Jiufotang Formation, temperate-subtropical swamps dominated by ginkgo and conifer trees, set among shallow lakes and stagnant waterways. **Size:** WS unknown; BL 17cm (7in); TL unknown **Features:** Head small & round. Teeth absent. Beak short. Legs relatively long w/ short tarsus. Wings & rectrices unknown. **Biology:** Very similar in overall anatomy to other confuciusornithids, but in some ways more similar to modern birds (Cau 2010a). For example, the short torso w/ fewer vertebrae than other species & long ischium may suggest that this species is a confuciusornithid-like primitive euornithean or enantiornithean. Stomach contents reveal that this species fed at least partially on fish.

Hengdaozi Great Wall Bird *Changchengornis hengdaoziensis*
Time: 122 Ma ago **Location:** Liaoning, China **Habitat:** Upper Yixian Formation (see above) **Size:** WS 42cm (1.4ft); BL 20cm (8in); TL >30cm (1ft) **Features:** Very small. Beak long & hooked. Crown prominently round-crested. Wings short & relatively broad. Legs short w/ partially reversed hallux. Ribbon-tailed w/ single pair of rectrices smaller than in related species. **Biology:** Possibly semi-arboreal, but the smaller, unspecialized wings suggest a poor flier. Probably foraged primarily on the ground.

Chaoyang SAPE Bird *Sapeornis chaoyangensis*
Time: 120 Ma ago **Location:** Liaoning, China **Habitat**: Jiufotang Formation **Size:** WS 1.4m (4.6ft); BL 45cm (1.5ft); TL unknown **Features:** Head high & round with narrow snout. Eyes large. Forward-pointing teeth restricted to upper jaw tip. Lower jaw toothless. Wings extremely long w/ large claws and long primaries (third primary longest, first shortest). Tail short. Legs short & stout. Toes w/ very large, strongly hooked talons. Hallux reversed. **Biology:** Fossils found in both terrestrial & lake sediments, indicating wide distribution (possibly primarily inhabiting the canopy). Adept perchers, they may have climbed up tree trunks w/ the large wing claws to escape predators & spent much of their time climbing among high branches. Incapable of strong flapping flight, the extensive wings may have been used for extended gliding/soaring between trees. Herbivorous or omnivorous, probably folivores & seed-eaters, supplementing w/ fruits & invertebrates. The only well-supported member of the *Omnivoropterygiformes*, "omnivorous wings", a bizarre early group of short-tailed birds, similar in some ways to the caenagnathiformes. Several additional species have been described, though they probably all represent growth stages of this single species.

Basal Enantiornitheans

The most diverse and successful group of Mesozoic birds were the enantiornitheans ("opposite birds"). They were named for the fact that their shoulder joint articulates in a fashion opposite that of the euornitheans ("true birds"). In modern birds, the shoulder blade (scapula) has a prominence that fits into a socket on the shoulder girdle (coracoid). This is reversed in enantiornitheans, in which the shoulder blade has the socket. This may indicate an independent evolution of strong, flapping flight.

Most enantiornitheans seem to have been land birds, dominating forested inland areas, and were correspondingly adapted for living in trees, with strongly reversed halluces, though some species may have been equally at home foraging on or fishing from the ground. Some species are known to have nested on the ground near rivers, lakes and other bodies of water in large colonies (Dyke & al, 2011).

Like most Mesozoic birds, enantiornitheans may have had growth patterns slightly different than typical modern birds. Studies of growth rings in their fossilized bones suggests that while they grew rapidly after hatching, their growth slowed significantly before reaching full size (Cambra-Moo & al., 2006). Some enantiornithean species are known to have rapidly developed their wings and were probably capable of flight from a relatively young age. In these species, the duration of parental care may have been relatively short. Because they did not reach full size in their first year like most modern birds, immature enantiornitheans may have had different roles, and occupied different ecological niches, than their parents, shifting diet and ecology as they matured, as has been suggested for more primitive birds and other theropods.

Due to their unique anatomy and global distribution, enantiornithean fossils are easily identified, but many have been named on the basis of extremely scrappy fossil remains, often only small pieces of bone. Thus many species, while undoubtedly unique, are of little scientific value. Additionally, the interrelationships of enantiornitheans are very poorly understood, with only one or two well-supported groups. The fragmentary nature of most species, the lack of ability to use phylogenetic bracketing due to poorly known relationships, and the sheer diversity of form in well-known species renders most enantiornitheans impossible to plausibly reconstruct. For that reason only well-known specimens, usually those with known skull material, are illustrated here.

Fengning Primitive Feather ***Protopteryx fengningensis***
Time: 131 Ma ago **Location:** Hebei, China **Habitat:** Sichakou Member, Huajiying Formation. Temperate woodland dominated by lakes. **Size:** WS 33cm (1ft); BL 16cm (6in); TL > 23cm (9in) **Features:** Head round w/ somewhat pointed snout, teeth restricted to tips of upper & lower jaws. Wings relatively short & broad (average primary feather length 45mm, longest remix 95mm). Single pair of long, thin rectrical ribbons. **Biology:** The earliest known birds definitely capable of powered flight. Short, broad wings would have been useful for maneuvering in densely forested environments, though the lack of a fan-tail and a primitively long alular digit would have made for clumsier flight than in modern arboreal birds. The flight apparatus was well-developed & the sternum keeled, but the wing retained separate bony digits and small claws.

Slender Near Protopter ***Paraprotopteryx gracilis***
Time: 124.6 Ma ago **Location:** Liaoning, China **Habitat:** Lower Yixian Formation. Temperate conifer and ginkgo forest set among a series of lakes fed by streams and runoff from a nearby range of active volcanic mountains. **Size:** (jubenile) WS >22cm (9in); BL >11cm (4in); TL >20cm (8in) **Features:** Known only from a subadult specimen. Wings short (average primary feather length 40mm). Hand fused at the base. Alular digit w/ small claw. Claw of major digit smaller than alular claw, & third digit claw vestigial. Legs relatively short w/ strongly clawed feet & reversed hallux. Tail short w/ four relatively short & thin rectrical ribbons. Each ribbon feather ends in an oval-shaped expansion. **Biology:** The wing anatomy lacks a procoracoid structure found in the similar *Protopteryx*, indicating a somewhat weaker flight ability. Like other primitive enantiornitheans, probably relied largely on the large wing claws to climb.

Leg-feathered Enantiornithean **No scientific name (specimen IVPP V13939)**
Time: 124.6 Ma ago **Location:** Liaoning, China **Habitat:** Lower Yixian Formation (see above) **Size:** WS 26cm (in); BL ~13cm; TL ~14cm **Features:** Wings large & long w/ short wing bones but long remiges (up to 6cm). Legs long, w/ long tarsus & slender toes w/ small claws. Alula very long. Feathers on legs up to 1.5 cm long, longer than some body feathers, but not forming a planar "hind wing" or extending onto the tarsus. Tail feathers very short forming a broad stub-tail. **Biology:** Slender legs, strongly hooked hallux claw, & large alula extending past the major digit probably means these birds were arboreal. It is possible that the unusually long leg feathers functioned in steering and maneuvering as in microraptorians.

Lake Volcano Bird ***Huoshanornis huji***
Time: 120 Ma ago **Location:** Liaoning, China **Habitat:** Jiufotang Formation (see above) **Size:** WS ~22cm; BL 9cm; TL unknown **Features:** Snout short & straight. Alular digit very small. Minor digit relatively long & linked to major digit, w/ significant space between the metacarpals. Claws on major & alular digits nearly equal in size. **Biology:** Unique wing arrangement & large inter-metacarpal space (anchor point for important feather-control muscles) suggests very high maneuverability at low speeds. Probably better than other enantiornitheans at controlling the spacing & alignment of their primary feathers & the shape of the wing tip, allowing precision flying. Only known fossil specimen bears what appears to be a well-preserved skull, but most of it is reconstructed and only the general shape of the snout can be established.

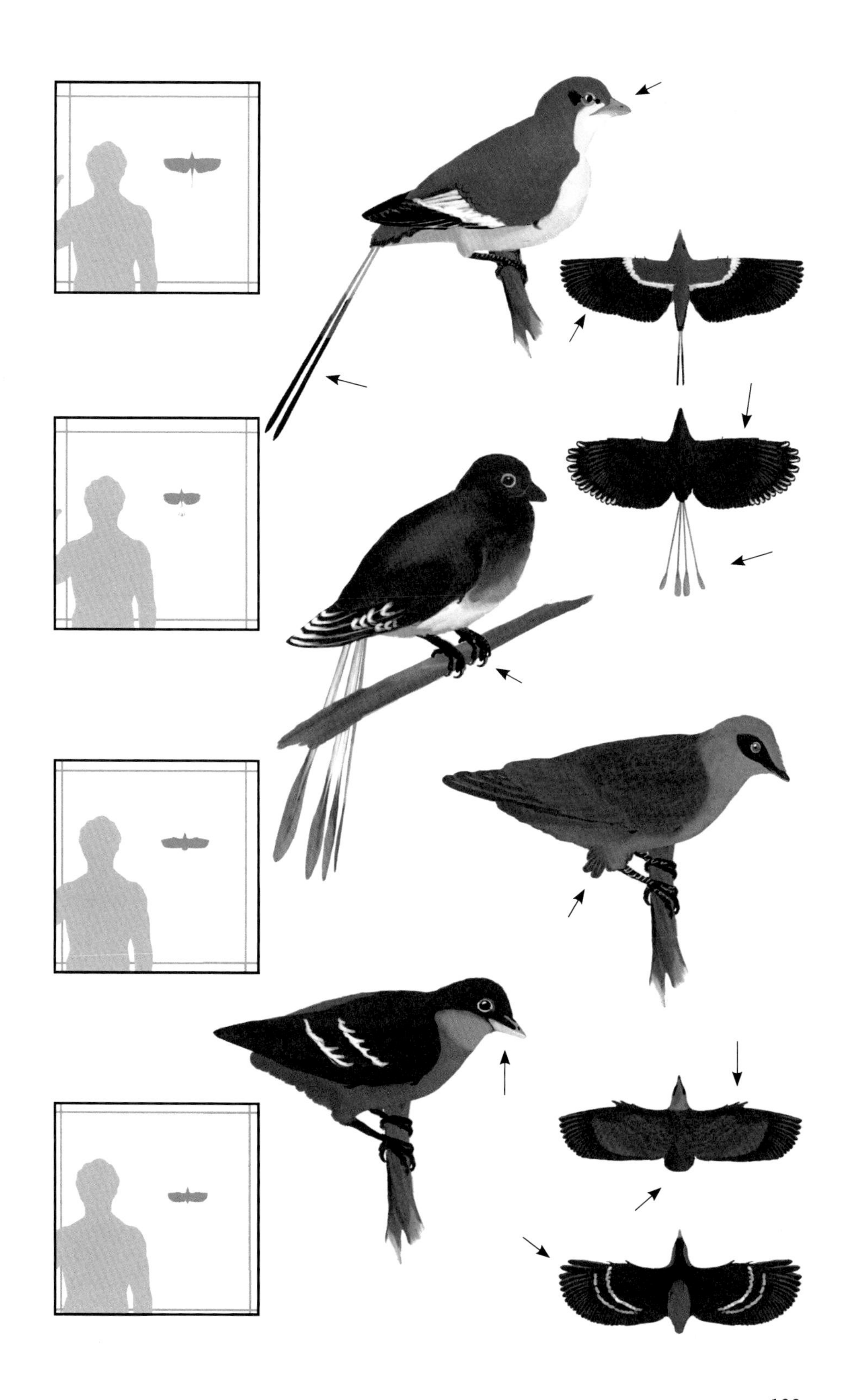

Meng's Shenzhou 7 Bird ***Shenqiornis mengi***
Time: 122 Ma ago **Location:** Hebei, China **Habitat:** Qiaotou Formation
Size: WS 40cm (1.3ft); BL 20cm (8in); TL 22cm (9in) **Features:** Relatively large head w/ long snout. Teeth large & conical but slightly recurved. Wings short (primary feather length 7cm) w/ unfused fingers & claws on the alular & major digits. Tail short w/ no evidence of rectrices. **Biology:** Probably arboreal & weak fliers. Like other primitive enantiornitheans, alular digit large & strongly clawed. Lacked procoracoid bones which normally indicate strong flapping flight ability. The unusually large, conical teeth & heavily built skull indicate that these may have fed on aquatic molluscs, arthropods, & other hard-shelled invertebrates.

Hebei Thin Bird ***Vescornis hebeiensis***
Time: 122 Ma ago **Location:** Hebei, China **Habitat:** Qiaotou Formation **Size:** WS 25cm (10in); BL 10cm (4in); TL 13cm (5in) **Features:** Snout small & rounded, containing teeth. Legs relatively long w/ large feet. Wings rounded but w/ long primary feathers relative to skeletal wing & hand length. Claws on alular & major digits, minor digit lacks claw. Keeled sternum. **Biology:** Small size & small snout w/ small teeth suggest insectivorous diet.

Luan Jibei Bird ***Jibeinia luanhera***
Time: 122 Ma ago **Location:** Hebei, China **Habitat:** Qiaotou Formation **Size:** WS ~25cm (10in); BL 10cm (4in); TL unknown **Features:** Only known specimen is poorly preserved & generally very similar to contemporary *Vescornis*. Differs in the presence of a minor digit claw, & lack of a keeled sternum. May be based on a juvenile specimen. **Biology:** Very similar to *Vescornis hebeiensis*, possibly representing the same species. The two are reconstructed here in such a way that they could be taken to represent an adult and juvenile of the same species.

Tiny Gobi Feather ***Gobipteryx minuta***
Time: 75-72 Ma ago **Location:** Omnogovi, Mongolia **Habitat:** Barun Goyot Formation & Ukhaa Tolgod, Djadochta Formation. High desert. Dune fields and arid scrubland. **Size:** WS >30cm (1ft); BL 17cm (7in); TL unkown **Features:** Bill large, broad, triangular & toothless, w/ rounded tips & covered in a keratinous beak. **Biology:** Several eggs & embryos of an enantiornithean from the same location may belong to *G. minuta*. Eggs short & oval-shaped.

Romeral's Iberian Middle Bird *Iberomesornis romerali*
Time: 125 Ma ago **Location:** Cuenca, Spain **Habitat:** La Huerguina Formation. Forested beaches surrounding the large, shallow-water lake Las Hoyas with high concentration of aquatic life. **Size:** WS >20cm (8in); BL 9cm (3.5in); TL unknown **Features:** Wings relatively short but hand & primary feather lengths unknown. Feet large w/ large, curved claws. **Biology:** The smallest known non-avian dinosaur species, weighing around 15-20 grams (Sanz & Ortega 2002).

Las Hoyas Dawn Bastard-wing Bird *Eoalulavis hoyasi*
Time: 125 Ma ago **Location:** Cuenca, Spain **Habitat:** La Huerguina Formation (see above) **Size:** WS >20cm (8in); BL ~9cm (3.5in); TL unknown **Features**: Wings long, w/ ulna over twice the length of carpus and significantly longer than humerus. Long, well developed primary feathers & alula. Sternum fish- or hourglass-shaped but narrow, possibly indicating a juvenile specimen. **Biology:** Stomach contents of the only known fossil specimen include fragments of crustacean exoskeletons, providing direct evidence of diet in this species. These birds probably foraged in or near the water at least some of the time, as shoreline generalists. The wishbone shape is consistent with a flap-gliding mode of flight.

Long-toed Liaoning Bird *Liaoningornis longidigitris*
Time: 124.6 Ma ago **Location:** Liaoning, China **Habitat:** Lower Yixian Formation. Temperate conifer and ginkgo forest set among a series of lakes fed by streams and runoff from a nearby range of active volcanic mountains. **Size:** WS unknown; BL ~18cm (7in); TL unknown **Features**: Breastbone broad, keeled, & well-developed, with an hourglass- or fish-like shape. Wings strong and robust but incompletely known. Tarsus short & stout with long toes. **Biology**: The breastbone, which is similar in general shape to, but much broader than, that of *Eoalulavis*, indicates a stronger flier. The unusual hourglass-like breastbone shape among enantiornitheans may be a juvenile characteristic (O'Connor 2010).

Graffin's Heaven Bird *Qiliania graffini*
Time: 120 Ma ago **Location:** Gansu, China **Habitat:** Xiagou Formation. Wide basin dominated by a system of large, tranquil freshwater lakes. **Size:** WS unknown; BL ~20cm (8in); TL unknown **Features:** Legs long & slender. First toe relatively large, fourth toe relatively slender, w/ an unusually small claw. Claws long & needle-like w/ very weak curvature. Internally, the hip bones are fused & the pubic bone tapers to a deflected point rather than an expanded boot. **Biology:** Based on the curvature of the toe claws, it is likely that these were equally likely to forage on the ground as in trees, inhabiting an ecological niche similar to doves or cuckoos.

Eoenantiornithiformes

"Dawn enantiorns" were relatively primitive opposite birds that may form a natural group including the short-snouted eoenantiornithids and the long-snouted longipterygids, "long-wings", one of the earliest known specialized groups of enantiornitheans. All seem to have been well adapted to perching, but somewhat paradoxically, many were primarily fishers and probers. Many likely inhabited Kingfisher-like niches, perching above lakes and rivers and swooping down to the surface to grab fish.

Buhler's Dawn Enantiorn ***Eoenantiornis buhleri***
Time: 124.6 Ma ago **Location:** Liaoning, China **Habitat:** Lower Yixian Formation. Temperate conifer and ginkgo forest set among a series of lakes fed by streams and runoff from a nearby range of active volcanic mountains. **Size:** WS 25cm (10in); BL 10cm (4in); TL 11cm (4.3in) **Features:** Head broad w/ short, deep snout, feathered to near the tip. Teeth larger in the front of the jaws. Wings short (distal primary 6.6cm; distal secondary ~6cm). Primaries anchored to a short hand fused at the base, rendering the alular digit unusually long relative to the rest of the hand. First two fingers bear large claws but the third finger is reduced & partially fused to the second. Feet small w/ reversed hallux. Tail short & rounded w/ no evidence of rectrices. **Biology:** Like *Protopteryx*, capable of powered flight coupled with several key primitive features (large alular digit, lack of aerodynamic rectrices) that would have made flight clumsy & landings imprecise. The retention of large wing claws suggests that they may have landed by clinging to large branches or tree trunks & then climbing to a perch, rather than alighting directly onto small branches.

Guo's Bohai Bird ***Bohaiornis guoi***
Time: 124.6 Ma ago **Location:** Liaoning, China **Habitat:** Lower Yixian Formation (see above) **Size:** WS 45cm (1.5ft); BL 20cm (8in); TL 36cm (1.2ft) **Features:** Head broad w/ short, deep snout. Teeth larger in the front of the jaws. Wings relatively long. Primary feathers (longest ~12cm) anchored to a robust hand w/ a long, clawed alular digit. Major digit claw reduced in size & minor digit lost. Legs relatively long, about as long as the wing skeleton. Feet broad w/ reversed hallux & large but relatively weakly curved claws. Tail short w/ two short, broad rectrical ribbons. **Biology:** Very similar to *E. buhleri* in overall anatomy, they differed primarily in larger size & teeth restricted more towards the tips of their jaws, as well as details of the internal skeletal anatomy. May represent mature form.

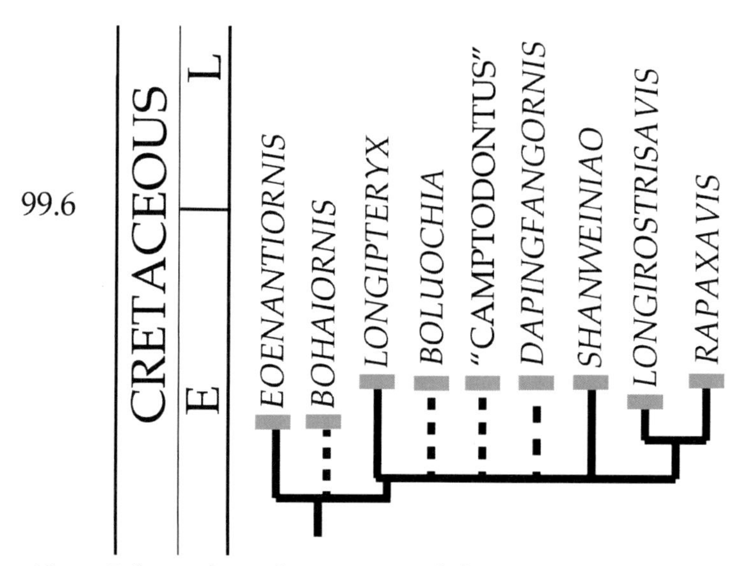

Above: Relationships of eoenantiornithiformes over time. Phylogeny approximated based on Cau & Arduini 2008 and O'Connor, Gau & Chiappe 2010.

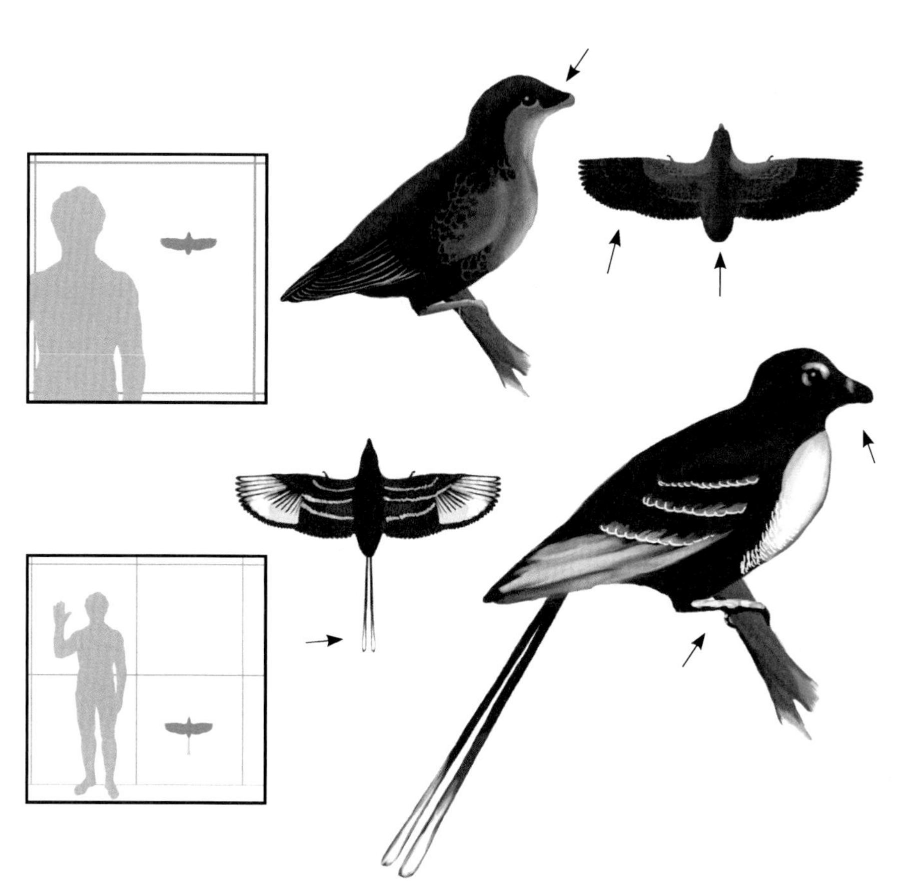

Thorn-nosed Dapingfang Bird ***Dapingfangornis sentisorhinus***
Time: 120 Ma ago **Location:** Liaoning, China **Habitat:** Jiufotang Formation, temperate-subtropical swamps dominated by ginkgo and conifer trees, set among shallow lakes and stagnant waterways. **Size:** WS 22cm (8.6in); BL 12cm (4.7in); TL 23cm (9in) **Features:** Head large w/ short, thin snout, feathered to near the tip. Teeth sharp and recurved. Feet w/ strong talons. Single pair of very long and thin wire-like rectrices tipped w/ broad vaned plumes. **Biology:** Probably carnivorous, feeding on fish, lizards & other small vertebrates. A supposed horn high on the snout may be an artifact of preservation, though it may reflect a pair of actual, laterally-projecting lachrymal bones (Mortimer 2010). However, these were shorter than the overlying feathers & probably would not have been visible in life.

Chaoyang Long-wing ***Longipteryx chaoyangensis***
Time: 120 Ma ago **Location:** Liaoning, China **Habitat:** Jiufotang Formation. Temperate-subtropical swamps dominated by ginkgo and conifer trees, set among shallow lakes and stagnant waterways. **Size:** WS 34cm (1.1ft); BL 16cm (6in); TL ~18cm (7in) **Features:** Snout very long w/ large, curved teeth restricted to the jaw tips. Wings large & broad (medial primary 9cm) w/ large alular & major claws. Minor claw vestigial. Legs short. Stub-tail lacked any long rectrices. **Biology:** Teeth large & conical but slightly flattened and curved; this, combined with their presence only in the jaw tips, has led to the suggestion that these were fishers. However, short leg and perching feet may indicate an arboreal, insectivorous lifestyle instead. Broad wings and wishbone anatomy suggest that these birds may have employed continuous flapping flight.

Zheng's Boluochi Bird ***Boluochia zhengi***
Time: 120 Ma ago **Location:** Liaoning, China **Habitat:** Jiufotang Formation (see above) **Size:** WS ~40cm (1.3ft); BL ~20cm (8in); TL unknown **Features:** Very similar to "C." *yangi*, also w/ a long, narrow snout and very large, strongly hooked teeth. Differed primarily in smaller size & foot anatomy, where the outer toe diverged significantly from the others. **Biology:** Possibly the same species as "C." *yangi*. *B. zhengi* was initially reported to have a hooked, raptorial beak, but this was a misinterpretation due to poor preservation and the unusually large, hooked teeth.

Yang's Bent-tooth **"Camptodontus"** ***yangi***
Time: 120 Ma ago **Location:** Liaoning, China **Habitat:** Jiufotang Formation (see above) **Size:** WS ~50cm (1.6ft); BL 25cm (10in); TL unknown **Features:** Head long w/ very long snout. Teeth extremely large & strongly curved w/ a distinctive bend at the tip, & restricted to the jaw tips. Wings long w/ alular & major claws. Legs relatively short. **Biology:** Similar in many respects to *Longipteryx*, but notable for the much larger teeth, which would have protruded conspicuously from the slender jaws in life. Larger than *Longipteryx*, these may also have fed on fish &/or arboreal prey, but possibly specializing in larger prey than *L. chaoyangensis*. The genus name *Camptodontus* is preoccupied and will need to be replaced if it is not a synonym of *Boluochia*.

Cooper's Fan-tail Bird ***Shanweiniao cooperorum***
Time: 122 Ma ago **Location:** Liaoning, China **Habitat:** Upper Yixian Formation (see above) **Size:** WS 32cm (1ft); BL 14cm (5in); TL >20cm (8in) **Features:** Head long w/ long, narrow snout, & teeth restricted to the jaw tips. Wings relatively long (distal primary 8.2cm), but hand bones reduced, & all digits lacked claws. Legs long w/ short tarsus & long toes w/ very large, gently curved claws. Up to six ribbon-like feathers formed the tail & overlapped each other at the base. **Biology:** The lack of wing claws indicates that these would have exclusively alighted on branches by grasping w/ the feet, rather than hooking foliage or tree trunks w/ the wing claws, as in some other enantiornitheans. The large number of tail feathers, which overlapped each other, may have evolved in parallel w/ the retractable fan-like tails of euornitheans. These would have granted higher maneuverability when flying among dense foliage & allowed more precise landings.

Pan's Grasping Bird ***Rapaxavis pani***
Time: 120 Ma ago **Location:** Liaoning, China **Habitat:** Jiufotang Formation (see above) **Size:** WS ~25 cm (10in); BL 17cm (7in); TL 30cm (11.8in) **Features:** Snout long, thin, & slightly down-curved, w/ small, slender teeth restricted to jaw tips. Wings short w/ short alular digit & lacking any claws. Legs relatively long, w/ long, forward-directed hallux. Foot claws extremely long, nearly the same length as the toes. Possible specimen has pair of rectrices w/ partly unbarbed vanes (O'Connor & al. 2012). **Biology:** Lack of a J-shaped 1st metatarsal may indicate a pamprodactyl foot (all four toes facing forward) as seen in modern swifts & mousebirds. This, combined w/ very large foot claws, long hallux, & lack of wing claws suggests arboreal perchers, alighting on branches from the air, rather than landing & climbing w/ assistance from the forelimbs. The toes would have been able to grasp twigs or even flat surfaces like tree trunks or rock faces in a pincer-like arrangement rather than the opposed grasping arrangement of normal reversed halluces. Long, thin snout w/ slender-toothed tip indicates a probing lifestyle, either foraging on the ground on in tree bark for invertebrates.

Yang's Shenyang Bird ***Shengjingornis yangi***
Time: 122 Ma ago **Location:** Liaoning, China **Habitat:** Jiufotang Formation (see above) **Size:** WS >22cm (8.6in); BL 20cm (7.8in); TL unknown **Features:** Snout relatively long & down-curved. Teeth large & conical, restricted to jaw tips. Upper wing bones equal in length, hands short & clawed. Leg slightly longer than skeletal wing. **Biology:** The combination of decurved snout & wing claws suggests an intermediate position between longirostrisavisines and other longipterygids. Feathers not preserved.

Han's Long-snout Bird ***Longirostravis hani***
Time: 122 Ma ago **Location:** Liaoning, China **Habitat:** Upper Yixian Formation (see above) **Size:** WS 32cm (1ft); BL 17cm (7in); TL >21cm (8in) **Features:** Head large w/ long, very thin & delicate snout. Teeth relatively long & peg-like, restricted to jaw tips. Wings long (primaries ~8cm) but broad & rounded (secondaries ~8cm, extending beyond bony tail). All digits lack claws. Legs & feet very small. Single pair of ribbon-like rectrices. **Biology:** Long thin snout resembles sandpipers, though anatomy of the feet shows that these were specialized perchers. May therefore have been arboreal insectivores &/or bark probers.

Cathayornithiformes

The "Cathay birds" probably include two groups. The 'cathayornithids' are a group of relatively primitive enantiornitheans which are united primarily by primitive characteristics (plesiomorphies). They may therefore not represent a natural group, but because they are superficially similar in appearance, they are placed together here for convenience. Some phylogenetic analyses have found these to represent a grade which gave rise to the avisauroids (Cau & Arduini 2008). Cathayornithids generally had wishbone anatomy consistent with gliding or flap-gliding flight.

The avisauroids ("bird lizards"), possible advanced members of the cathayornithiformes, were likely specialized perching birds, well adapted to life in the trees. These were probably generally predatory and similar to modern raptors in behavior. As the name suggests, avisauroids were originally thought to be close relatives of the deinonychosaurs rather than enantiornitheans, and even some recent analyses have yielded this result (Kurochkin & al. 2011), but those have been criticized by other researchers (Cau 2011). Only some primitive avisauroids are known from fossils preserving the skull. While one of these (*Cuspirostrirornis houi*) had a somewhat pointed snout, the premaxilla was full of teeth and there is no evidence that this or any other enantiornithien species, save *Gobipteryx minuta,* had beaks.

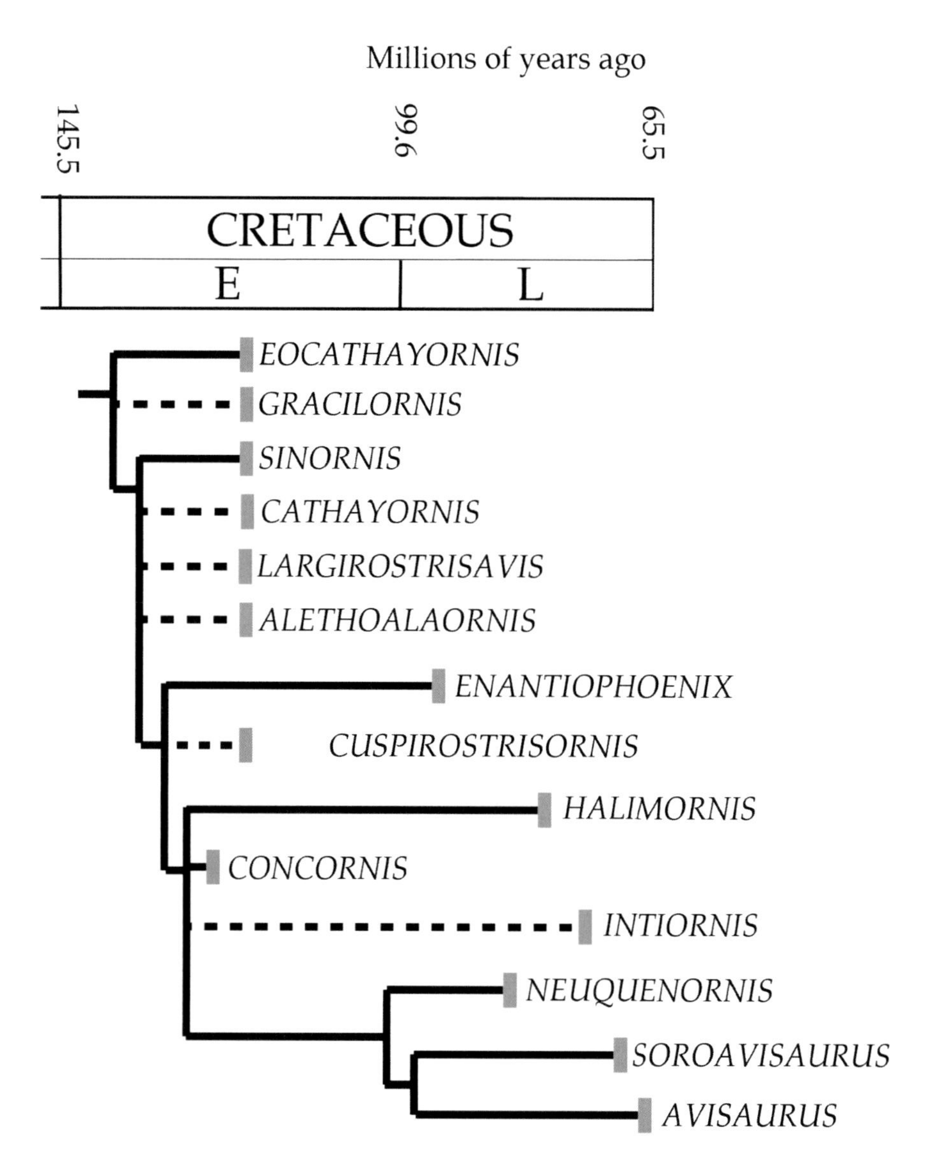

Above: Relationships of cathayornithiformes over time. Phylogeny approximated based on Cau & Arduini 2008.

Hou's Peng Bird ***Pengornis houi***

Time: 120 Ma ago **Location:** Liaoning, China **Habitat:** Jiufotang Formation (see above) **Size:** WS ~50cm (1.6ft); BL 25cm (10in); TL unknown **Features:** Head triangular. Teeth small & short, numerous & blunt. Wings short & broad w/ a strong alular claw. **Biology:** Small, blunt, "onion shaped" teeth only weakly curved & with signs of wear, suggesting a diet of arthropods, mollusks, & other shelled invertebrates. Probably employed continuous flapping flight.

Six-toothed Large-snout Bird ***Largirostrisornis sexdentoris***

Time: 120 Ma ago **Location:** Liaoning, China **Habitat:** Jiufotang Formation (see above) **Size:** WS >25cm (10in); BL 15cm (6in); TL unknown **Features:** Head small w/ relatively long & thin snout w/ six pairs of small, curved teeth in both the upper jaw & lower jaws. Body notably large & long compared w/ other enantiornitheans. Wings broad & rounded w/ short primary feathers. Legs short w/ large, strong perching claws. **Biology:** The breastbones of these birds supported a short but strong keel, & the hand bones were fused to a degree seen in some modern perching birds, suggesting they were capable fliers. The wings retained claws, however, suggesting that climbing among branches was still part of their ecology.

Yandica Cathay Bird ***Cathayornis yandica***

Time: 120 Ma ago **Location:** Liaoning, China **Habitat:** Jiufotang Formation (see above) **Size:** WS ~20cm (8in); BL 13cm (5in); TL unknown **Features:** Snout long, broad & slightly flattened at the tip. Head triangular in profile. At least four teeth in the upper jaw & three in the lower, possibly more. Wing short & rounded w/ short primary feathers. Alular digit straight w/ small claw. Small claw also present on the major digit. Legs long, w/ long tarsus & weakly curved claws. **Biology:** The head & teeth were rather primitive, similar to basal avialans like *Archaeopteryx lithographica*, and so these probably had a similar diet (likely insects & small vertebrates). The small wing claws, long legs & toes, & weakly curved claws may suggest a reduced role in climbing and perching, & while they were still undoubtedly arboreal (Bell & Chiappe 2010), they may have been more prone to ground foraging than other enantiornitheans. The broad, somewhat flattened snout tip may indicate that they preferred to dabble at lake or river shores.

Unusual Cathay Bird ***Cathayornis aberransis***

Time: 120 Ma ago **Location:** Liaoning, China **Habitat:** Jiufotang Formation (see above) **Size:** WS ~20cm (8in); BL 13cm (5in); TL unknown **Features:** Similar in most respects to *C. yandica*, & may differ only in very minor skeletal characteristics, if at all. May have had more teeth than *C. yandica*. **Biology:** Initially thought to have a small crest along the mid-line of the skull (possibly supporting a soft-tissue cockscomb or other structure in life), this interpretation was later found to be in error (O'Connor & Dyke 2010). However, such structures do not always correlate with the underlying bone, so a soft-tissue crest is retained here as a link to the historical interpretation & as a speculative distinguishing feature from *C. yandia*.

Chabu Cathay Bird **"Cathayornis" *chabuensis***
Time: 120 Ma ago **Location:** Inner Mongolia, China **Habitat:** Jingchuan Formation **Size:** WS ~33cm (1.1ft); BL 15cm (6in); TL unknown **Features:** Similar in outward appearance to *C. yandica* & other primitive enantiornithean species, but somewhat larger. **Biology:** Despite the superficial similarity to cathayornithids, these may not be closely related to *C. yandica*. The skull is unknown, and therefore so are their diet & habits, but they probably occupied a different ecological niche than the contemporary *Cathayornis* and *Sinornis*.

China Bird ***Sinornis santensis***
Time: 120 Ma ago **Location:** Liaoning, China **Habitat:** Jiufotang Formation (see above) **Size:** WS ~35cm (1.1ft); BL 17cm (7in); TL unknown **Features:** Snout long & broad. Head triangular in profile. Jaws fully toothed. Wing short & rounded w/ short primaries. Alular digit long & curved at the base, w/ small claw. Larger claw present on the major digit. Tarsus shorter than the third toe + claw. Internally, tail short compared with *Cathayornis*, & hips broader. **Biology:** The longer, more hooked, & larger-clawed wing digits may imply a slightly greater degree of arboreality compared with *Cathayornis*. Short tarsus, long toes & strongly hooked talons may additionally imply a carnivorous diet.

Walker's Dawn Cathay Bird ***Eocathayornis walkeri***
Time: 120 Ma ago **Location:** Liaoning, China **Habitat:** Jiufotang Formation (see above) **Size:** WS ~30cm (1ft); BL 13cm (5in); TL unknown **Features:** Snout short & triangular. Wings short & stout but powerfully built. Three wing claws present, but small. Legs relatively long. **Biology:** The stout wing bones, small minor digit claw, & small but strong and keeled breastbone indicate strong flight ability.

Jiufotang Slender Bird ***Gracilornis jiufotangensis***
Time: 120 Ma ago **Location:** Liaoning, China **Habitat:** Jiufotang Formation (see above) **Size:** WS ~30cm (1ft); BL 13cm (5in); TL unknown **Features:** Snout short & triangular. Wings short. Legs relatively long. Overall more slender in build than *Eocathayornis*. **Biology:** Possibly related to the *Eocathayornis* and may have been similar in habits. A halo of long feathers on the skull & neck in the only known fossil resemble a crest or crown, & while this may well be due to decomposition, it is used here as a speculative distinguishing feature.

Lake Cuenca Bird *Concornis lacustris*
Time: 125 Ma ago **Location**: Cuenca, Spain **Habitat:** La Huerguina Formation. Forested beaches surrounding the large, shallow-water lake Las Hoyas with high concentration of aquatic life. **Size:** WS 22cm (8.6in); BL ~9cm (3.5in); TL ~10cm (4in) **Features:** Wings long w/ short skeletal wing, but long primary feathers. Small claws present on alular & major digits. Legs long & slender w/ large, hooked claws, & an especially large hallux claw. Breastbone lacked a keel on its front half. **Biology:** Unlike *C. houi*, these had proportionally short forelimbs w/ long primary feathers, possibly indicating greater reliance on flapping flight, as opposed to soaring, though the wishbone anatomy suggests a highly unique form of flight unlike any modern birds. Lakeshore habitat & long legs may indicate a specialization in small aquatic prey, though the skull is not known.

Hou's Point-snout Bird *Cuspirostriornis houi*
Time: 120 Ma ago **Location:** Liaoning, China **Habitat:** Jiufotang Formation (see above) **Size:** WS 24cm (9.5in); BL 10cm (4in); TL ~12cm (4.7in) **Features:** Snout short w/ a pointed tip, but lacking a beak. Five pairs of teeth each in upper & lower jaw. Wings short & rounded w/ relatively short primary feathers (~4cm long). Legs long w/ long toes & very large talons. **Biology:** Like other avisauroids, these may have been similar to modern raptors in appearance & habits. The long legs w/ short metatarsus & large talons suggest they were employed in prey-grappling behavior. Similarly, the long wing bones w/ proportionally short primary feathers are consistent w/ a soaring/flapping flight style.

Amber-loving Opposite Phoenix *Enantiophoenix electrophyla*
Time: 95 Ma ago **Location:** Mount Lebanon, Lebanon **Habitat:** Ouadi al Gabour Formation **Size:** WS unknown; BL ~20cm (8in); TL unknown **Features:** Known from a fragmentary skeleton distinguished by internal features of the shoulder blade. Legs relatively stout, w/ large, short claws. **Biology**: Only known specimen contains small pieces of amber near the stomach region, suggesting these may have been arboreal sap-eaters.

Thompson's Seabird *Halimornis thompsoni*
Time: 80 Ma ago **Location:** Alabama, USA **Habitat:** Mooreville Chalk Formation. Shallow marine environment near the southeastern coast of the Western Interior Seaway. **Size:** WS ~40cm (1.3ft); BL ~17cm (7in); TL unknown **Features:** Small avisauroids characterized by unique features of the wing & leg bones (crest of the humerus nearly at the point of shoulder contact, femur expanded near the far end). **Biology:** These appear to have been marine, as the fossils were found nearly 50km from where the nearest shoreline would have been at the time. Expansion at the knee joint may indicate some unique adaptations of the leg, possibly for swimming. While the skull is unknown, they likely fed on fish & possibly other small birds.

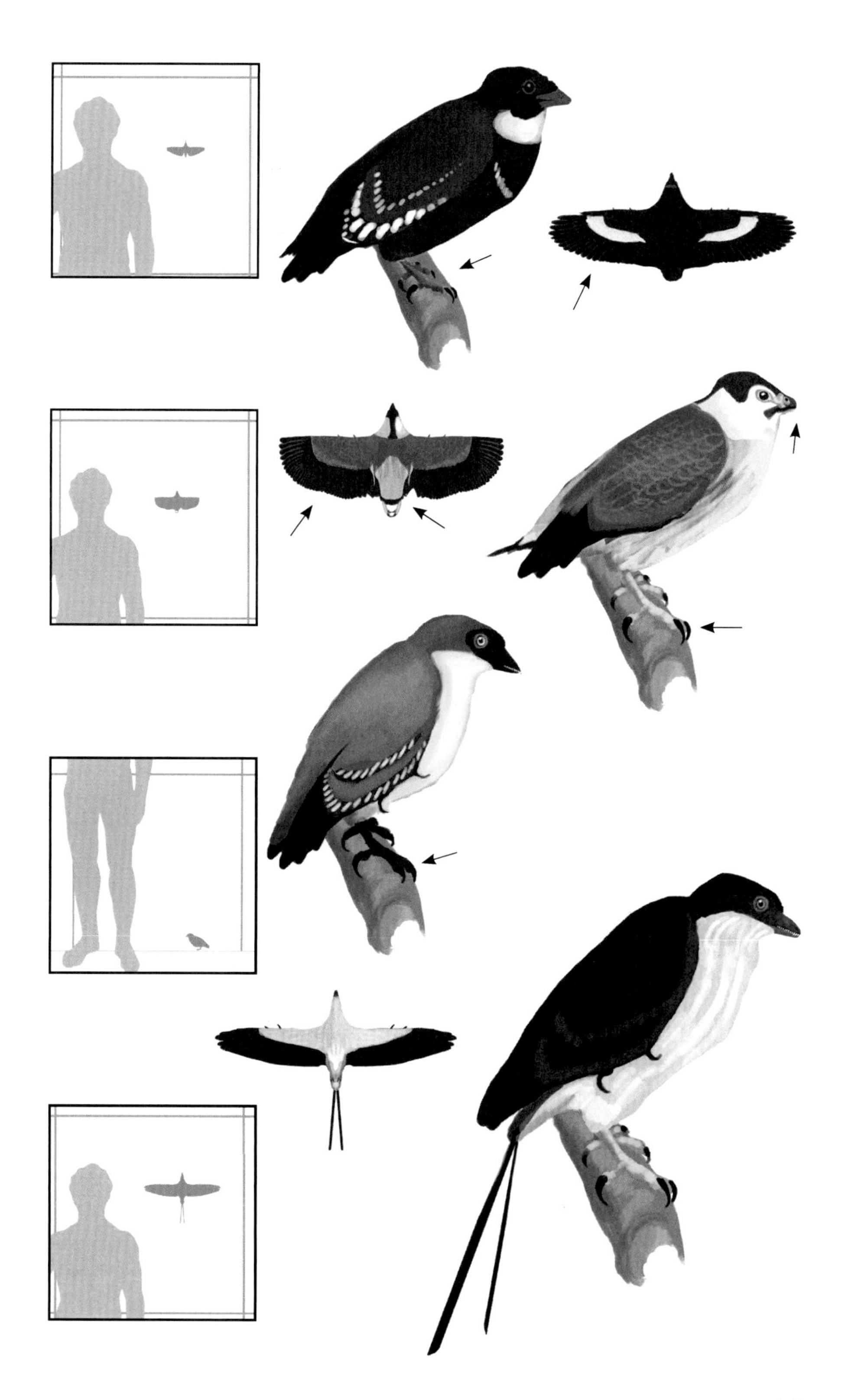

Unexpected Sun Bird ***Intiornis inexpectatus***

Time: 75 Ma ago **Location:** Salta, Argentina **Habitat:** Las Curtiembres Formation. Warm, semi-tropical forests near open plains dominated by marshes and rivers. **Size:** WS unknown; BL ~15cm (6in); TL unknown **Features:** Known from lower leg & foot. Tarsus long & slender. Toes long & slender, w/ large, highly curved talons. Three primary toes nearly equal in length. Second toe pointed inward. Hallux large but short & stocky w/ an especially large claw. **Biology:** The long, equal-length toes & large, curved talons are consistent w/ a specialized perching lifestyle.

Flying Neuquen Bird ***Neuquenornis volans***

Time: 85 Ma ago **Location:** Neuquen, Argentina **Habitat:** Bajo de la Carpa Formation. Braided, slow-flowing streams on open plains on the margins of more densely forested areas. **Size:** WS >30cm (1ft); BL ~15cm (6in); TL unknown **Features:** Wings very long. Legs relatively long w/ large, strongly hooked talons. **Biology:** The combination of long, relatively slender legs & large, hooked talons suitable for perching would have allowed acquisition of prey both from a perch & via ground or shallow-water foraging.

Southern Sister Avisaur ***Soroavisaurus australis***

Time: 70 Ma ago **Location:** Salta, Argentina **Habitat:** Lecho Formation **Size:** WS unknown; BL ~40cm (1.3ft); TL unknown **Features:** Known only from tarsus & foot bones. Tarsus short & stout. Hallux claw proportionally small compared to other avisaurids. Second toe pointed inward rather than forward. **Biology:** The large size & strong talons suggest these were perching & probably raptorial. They may have had an ecological niche similar to some modern raptors, spotting prey at a distance (either in fields or in rivers and lakes). The genus may be a synonym of *Martinavis*. (Walker & al. 2007).

Archibald's Bird Lizard ***Avisaurus archibaldi***

Time: 65.5 Ma ago **Location:** Montana, USA **Habitat:** Hell Creek Formation. Forested near-coastal flood plains dominated by flowering shrub species and coniferous trees. **Size:** WS unknown; BL ~45cm (~1.5ft); TL unknown **Features:** Known only from foot bones (tarsometatarsus) & an undescribed partial skeleton. Tarsus short & stout. The second toe was pointed inward rather than forward. **Biology:** The inward-pointing toe & short, robust tarsus in some ways resemble those of eudromaeosaurs. These may have had a raptorial function in both groups, used for pinning and seizing prey, making avisaurs possibly equivalent to modern birds of prey in ecology.

Cyril's Mystery Bird *Mystiornis cyrili*
Time: 125 Ma ago **Location:** Kemerovo, Russia **Habitat:** Shestakovo Formation **Size:** WS unknown; BL ~20cm (8in); TL unknown **Features:** Thrush-sized enantiornitheans known only from a fully-fused tarsometatarsus. Distinctive in forming a long, slender, arched tarsus w/ the second toe anchored much farther up the leg than the third & fourth, close to the hallux. **Biology:** The wide, thin tarsus, w/ high second toe, is typical of specialized swimming birds (see also euornitheans such as *Gansus*). Despite having been described as a primitive bird of a unique "order", it is more likely a specialized avisauroid (Cau 2011, Holtz 2011).

Enantiornithiformes

These enantiornitheans (which may or may not all form a natural group) showed a diverse array of lifestyles, including waders and divers. They seem to have been shorebirds with at least some species inhabiting inland, freshwater habitats on the banks of rivers. Some nested on the shore in huge colonies, where eggs were deposited in simple scrapes.

Short-footed Yunga Bird *Yungavolucris brevipedalis*
Time: 70 Ma ago **Location:** Salta, Argentina **Habitat:** Lecho Formation
Size: WS unknown; BL ~35cm (1.1ft); TL unknown **Features:** Known only from metatarsals & possibly some wing bones. Tarsus very unusual in being short & flared widely at the end. **Biology:** The squat, flattened & flared-out tarsus may indicate an aquatic, diving or swimming lifestyle, making these very unique compared the the other species of enantiornithiformes present in the Lecho ecosystem. *Elbretornis bonapartei*, known only from wing bones consistent in size with *Yungavolucris* tarsi, may be a synonym (Mortimer 2010). The *Elbretornis* specimen consists only of a partial humerus & shoulder girdle, however, so a wingspan estimate & in-flight reconstruction cannot be completed.

El Brete Lecho Bird *Lectavis bretincola*
Time: 70 Ma ago **Location:** Salta, Argentina **Habitat:** Lecho Formation **Size:** WS unknown; TL ~60cm (2ft); BL unknown **Features:** Lower leg long & thin, w/ distinctive forward-projecting ankle joint. Tarsus broader but very thin in profile & relatively long. **Biology:** Known from lower leg (tibiotarsus) & tarsal bones (tarsometatarsus), may be a synonym of *Enantiornis leali* or *Elbretornis bonapartei* (Walker & Dyke 2010).

Basal Euornitheans

"True birds", the euornitheans include all birds more closely related to those living today than to opposite birds. All known euornitheans had fan tails. In contrast to the long, narrow, separated display feathers of the ribbon-tailed birds, these had an array of vaned feathers attached to a true pygostyle that allowed the feathers to expand into an overlapping, aerodynamic fan that greatly improved maneuverability, allowing for precise landings and more intricately controlled flight.

Also unlike enantiornitheans, all known euornitheans had beaks homologous with those of modern birds. However, most Mesozoic fan-tailed birds retained teeth in their jaws, with beaks restricted to the jaw tips (the opposite of the condition in enantiornitheans, many of which had teeth restricted to the tips of narrow snouts), and often, if not always, supported by a unique predentary bone in the lower jaw which was lost in modern birds. The beaks of most early euornitheans were compound, made up of several discrete plates of keratin rather than singular solid structures (this condition persists in some modern birds, e.g. albatross). Some euornitheans experimented evolutionarily with toothlessness in various forms, but only the modern birds lost teeth altogether in favor of large beaks and bills.

A semi-aquatic lifestyle was probably primitive for this group, as evidenced by their toothed jaws with beaked tips, association with aquatic or shoreline environments, and evidence that many of them fed on fish or other aquatic animals. During the Mesozoic, forest and inland environments seem to have been dominated almost exclusively by enantiornitheans, while aquatic and shoreline environments were ruled by euornitheans. Fan-tailed birds seem to have made few inroads to inland ecosystems until the extinction of the enantiornitheans at the end of the Cretaceous period. However, their shoreline habitat led the euornitheans to evolve more highly developed shoulder girdles which allowed them to take off from a flat surface (Martin & al. 2012). Enantiornitheans likely needed to climb into branches or up tree trunks to become airborne, necessitating the retention of the wing claws in most species. Euornitheans, no longer reliant on climbing to achieve flight, reduced their wing claws to vestiges relatively early in their evolution.

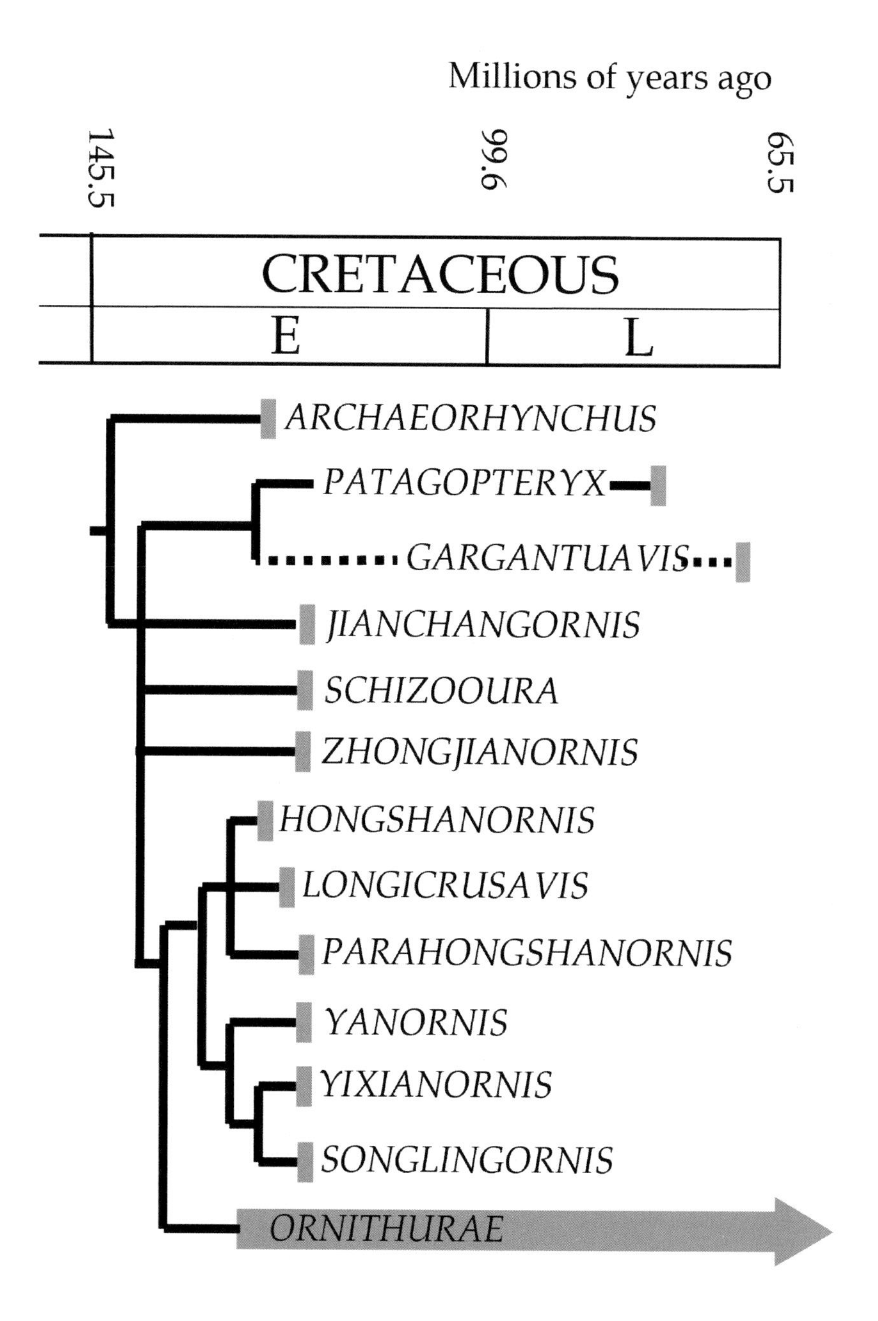

Above: Relationships of euornitheans over time. Phylogeny approximated based on You & al. 2006 and O'Connor & Zhou 2012.

Spatulate Ancient Beak ***Archaeorhynchus spathula***
Time: ~125 Ma ago **Location:** Liaoning, China **Habitat:** Yixian Formation **Size:** WS 40cm (1.3ft); BL 17cm (6in); TL 21cm (8in) **Features:** Head small & round w/ thin, toothless jaws tipped w/ flat beak. Neck short. Wings very long. Ulna ~ ¾ torso length. Humerus slightly shorter. Manus length unknown but very long. Primaries up to 10cm long, w/ long secondaries & rounded wing. Small claw on at least major digit. Legs short and stocky. Femur and tibia each ~ ¾ ulna length. Tarsus short, less than half tibia length. Toes long w/ small, weak claws. Rectrices long (up to 4cm) & arranged in a fan. **Biology:** Terrestrial, but probably not specialized waders. Small spoon-shaped bill indicates semi-aquatic, possibly surface-swimming lifestyle. However, the round bill w/o pointed tip supported by a predentary rules out a specialized ability for catching fish or other small aquatic animals. Instead, presence of gizzard stones may imply a primarily herbivorous diet of water plants, larvae, etc.

Long-crested Hongshan Bird ***Hongshanornis longicresta***
Time: 124.6 Ma ago **Location:** Liaoning, China **Habitat:** Lower Yixian Formation **Size:** WS 35cm (1.1ft); BL 13cm (5in); TL 18cm (7in) **Features:** Head small & round, long feathers possibly forming large crest. Snout short & slender w/ beaked tip. Few very small teeth behind beak in upper & possibly lower jaws. Wings very long, narrow & pointed, but w/ small forelimb (humerus, ulna & manus ~ same length as skull). Large claws on alular and major digits, minor digit fused to major digit. Legs very long & slender, w/ very long tibia & tarsus longer than toes. Hallux small & non-grasping. Toes long & slender, w/ small, weakly curved claws. Tail fan w/ at least four rectrices at ~5cm long. **Biology:** Likely wading birds due to long tibia & tarsus. Long legs allowed excursions into moderately deep areas of marshes or lake margins. Long narrow wings typical of aquatic birds; this, combined with the wishbone shape, is consistent with a dynamic soaring flight style. Straight, pointed beak, with small teeth would have made these less efficient at catching fish than toothed forms. Instead, bill may have been an adaptation for aquatic insectivory or mud-probing.

Hou's Long-shank Bird ***Longicrusavis houi***
Time: 122 Ma ago **Location:** Liaoning, China **Habitat:** Upper Yixian Formation **Size:** WS 40cm (1.3ft); BL 15cm (6in); TL ~20cm (8in) **Features:** Similar to *Hongshanornis* but w/ slightly larger wings and more robust snout. Wing & leg proportions nearly identical, but w/ slightly longer humerus & more robust toes/claws. Differed in aspects of breastbone. **Biology:** Similar in ecology to earlier species *H. longicresta*. Some place these together in a clade, *Hongshanornithidae* (w/ possibly more advanced *Parahongshanornis*). *L. houi* may be descendant species of *H. longicresta*.

Chaoyang Near Hongshan Bird ***Parahongshanornis chaoyangensis***
Time: 120 Ma ago **Location:** Liaoning, China **Habitat:** Jiufotang Formation. Temperate-subtropical swamps dominated by ginkgo and conifer trees, set among shallow lakes and stagnant waterways. **Size:** WS >20 cm (8in); BL ~12cm (5in); TL unknown **Features:** Head unknown. Neck short. Wings shorter relative to legs than relatives. All major wing segments about the same length. Alular digit short & robust w/ large claw. Smaller, thinner claw on major digit. Legs long w/ long tibia, tarsus longer than toes. Minor digit short & fused to major. Toes thin w/ short, slightly curved claws. Hal-

lux very small. **Biology:** Some internal anatomical features, such as a thick, U-shaped wishbone, led researchers to initially classify this species among the hongshanornithids. However, some preliminary cladistic analyses suggest these may have been more advanced euornitheans instead (Cau 2012).

Yang's Zhongjian Bird ***Zhongjianornis yangi***
Time: 120 Ma ago **Location:** Liaoning, China **Habitat:** Jiufotang Formation (see above) **Size:** WS ~75cm (2.5ft); BL 30cm (1ft); TL unknown **Features:** Head small relative to body. Toothless beak short but slender & pointed. Wings large but w/ very small claws. Legs long compared to the similar confuciusornithids. Large, highly curved claws on the feet. Tail short. **Biology:** The presence of toothless beaks in this species & the overall similarity to *Schizooura* may suggest a close relationship between the two. The anatomy of the feet & their large claws imply an ability to perch, & the lack of large wing claws suggests that they were adept at launching from the ground.

Li's Split-tail ***Schizooura lii***
Time: 120 Ma ago **Location:** Liaoning, China **Habitat:** Jiufotang Formation **Size:** WS 65cm (2ft); BL 30cm (1ft); TL 55cm (1.8ft) **Features:** Head small & triangular w/ slender pointed bill. Jaws toothless & probably beaked. Neck slightly longer than skull. Wings large, w/ ulna & manus slightly longer than humerus (ulna over half the length of the torso). Very long primary & secondary feathers resulting in broad, rounded wings. Small claw on alular digit. Major digit "claw" very small, blunt, & probably internal. Minor digit fused to major. Legs long, w/ tibia longer than femur & tarsus just over half tibia length. Toes relatively short & robust w/ robust but weakly curved claws. Fan-tail long, broad & triangular when expanded, with prominent fork. **Biology:** Probably ground foragers, as suggested by foot anatomy. Lack of teeth acquired independently of modern birds and possibly, independently of other primitive euornitheans like *Archaeorhynchus*. Whereas some early euornitheans retained teeth in addition to a premaxillary & predentary beak, possibly to aid in catching & holding fish, the loss of teeth in species like this may indicate a non-piscivorous diet consisting of more grains or other plant material, &/or arthropods. The broad, forked tail would have been detrimental (or at least neutral) to flight ability & was probably used in mating displays. The strong, rounded wings are likely adaptations to flying in a densely forested environment, & possibly to help compensate for the unwieldy tail feathers.

Small-toothed Jianchang Bird ***Jianchangornis microdonta***
Time: 120 Ma ago **Location:** Liaoning, China **Habitat:** Jiufotang Formation **Size:** WS ~60cm (ft); BL 34cm (1.1ft); TL unknown **Features:** Head triangular. Numerous small conical teeth in mid to rear jaws. Jaw tips beaked. Wings long w/ strong muscles anchored to keeled breastbone. Alular digit very long & strongly clawed. Major & minor digits bore very small claws. Minor digit fused to major. Feathers poorly preserved, but primary feathers over 15cm long. Legs short, w/ short tarsus ~equal to toes in length. Feet small w/ small, weakly curved claws. **Biology:** Evidence indicates these ate small fish including *Jainichthys*, & so probably foraged in or near lakes, though the relatively short tarsus & average-length tibia argue against a wading lifestyle. May have been surface swimmers or beachcombers.

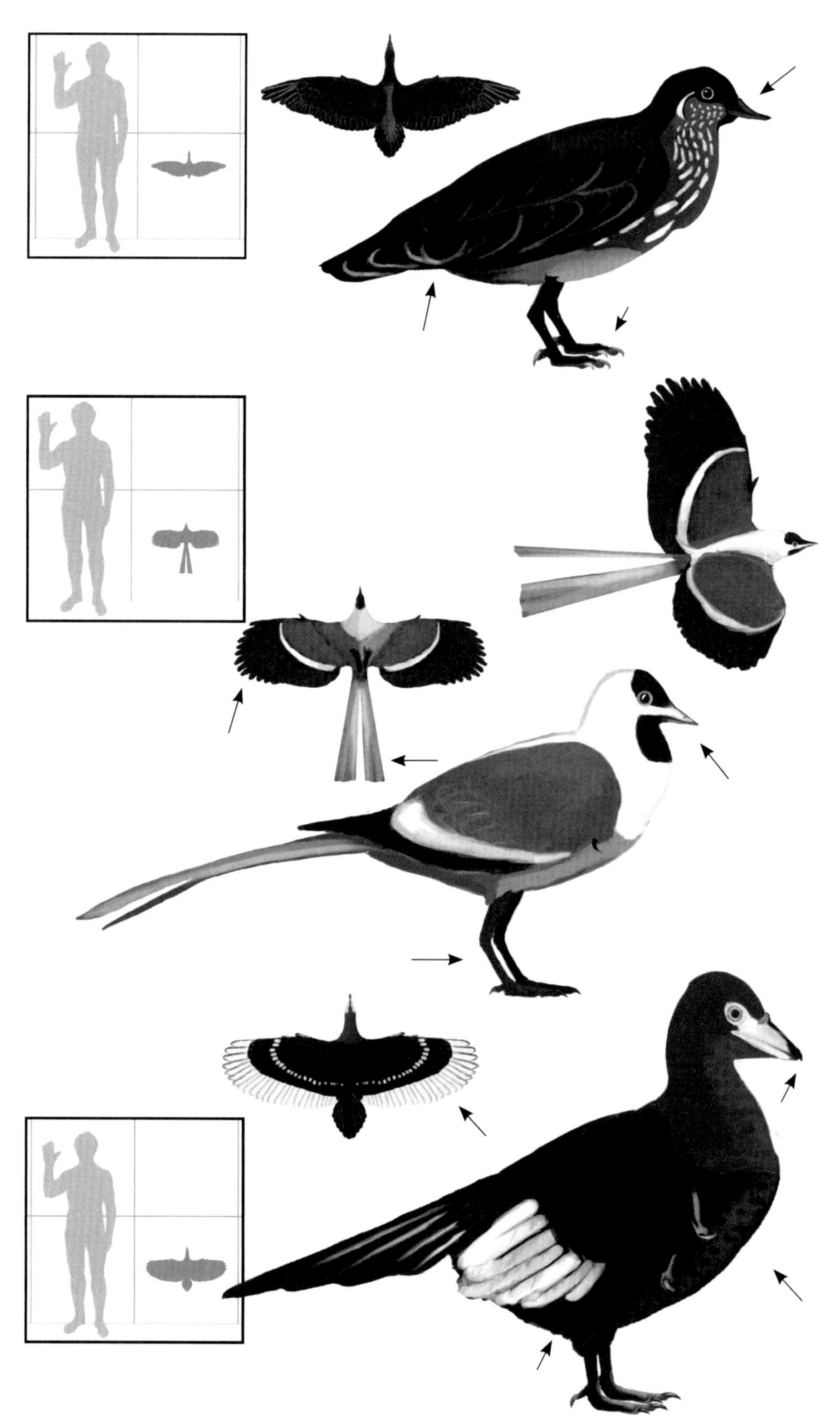

Martin's Yan Bird ***Yanornis martini***
Time: 120 Ma ago **Location:** Liaoning, China
Habitat: Jiufotang Formation **Size:** WS ~80cm (2.6ft); BL 32cm (1ft); TL unknown **Features:** Head small & triangular, w/ small beak at jaw tips. Teeth located behind beak. Neck long. Wings long (manus, ulna & humerus ~equal in length, ¾ the length of torso), broad, and powerful. Alular digit clawed. Small claw on major digit. Minor digit fused to major. Legs short, w/ femur & tibia equal in length, tarsus ¾ length of tibia. Toes equal to tarsus length w/ robust but weakly curved claws. Hallux small. **Biology:** Evidence shows that these fed on small fish, but also ingested large amounts of gizzard stones like herbivorous birds, so were probably omnivorous generalists. May have employed seasonal diet switching. Feet characteristic of ground foragers. Robust body, small head & long neck may have given these an appearance similar to modern game birds like pheasants.

Grabou's Yixian Bird ***Yixianornis graboui***
Time: 120 Ma ago **Location:** Liaoning, China **Habitat:** Jiufotang Formation **Size:** WS 60cm (2ft); BL 27cm (11in); TL 35cm (1.1ft) **Features:** Head small & triangular, w/ small beak at jaw tips. Teeth short & peg-like. Neck long. Wings short (humerus & ulna just over half torso length, manus shorter than ulna) but broad (primary feather length up to 15cm). Alular & major digits w/ small weakly curved claws. Minor digit fused to major. Legs long w/ long tibia but short tarsus (less than half tibia length). Toes very long & thin (nearly twice the length of the tarsus). Toes w/ small & very weakly curved claws. Tail w/ at least eight pennaceous rectrices up to 8cm long arranged into expandable fan, somewhat rounded in shape when expanded. **Biology:** Similar to *Y. martini*, but w/ much longer legs, longer, more slender toes, & somewhat shorter wings. Comparison of leg & wing proportions to modern birds indicate these may have been semi-aquatic, foraging both from the water surface & on the ground like modern ducks & geese. May have employed some degree of foot-propelled diving (Bell & Chiappe 2010). While there is currently no evidence of webbed feet, the long, slender toes & small, nearly straight claws would be consistent w/ this ecology.

Linghe Songling Bird ***Songlingornis linghensis***
Time: 120 Ma ago **Location:** Liaoning, China **Habitat:** Jiufotang Formation **Size:** WS unknown; BL 10cm (4in); TL unknown **Features:** Head slender w/ thin, pointed jaws & numerous, close-packed teeth. Jaw tips beaked. Wings known from fragments, but indicate well-developed shoulder girdle musculature. **Biology:** Similar in many respects to other yanornithiformes, except for small size & better-developed wings. Probably omnivorous shorebirds wading or swimming after fish & other forage.

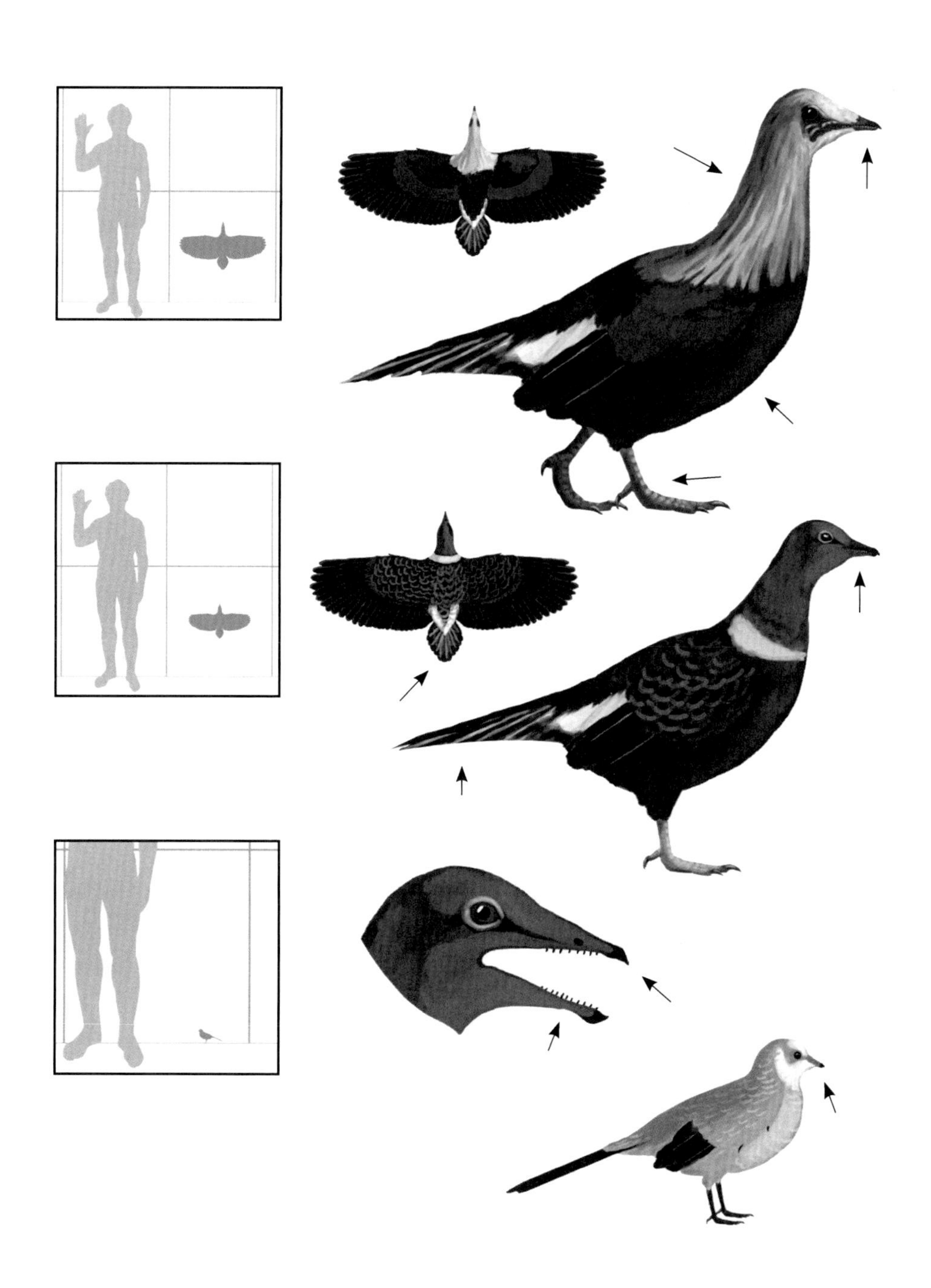

Patagopterygiformes

"Patagonia wings" are enigmatic birds, all poorly known from fragmentary material, which may or may not form a natural group. Several of them appear to have been flightless ground birds, and they may have descended from a late-surviving lineage of primitive euornitheans or ambiortiformes. The poorly known *Alamitornis minutus* from Argentina may be an additional member, but is too incompletely preserved to restore here.

Patagonia Wing ***Patagopteryx deferrariisi***
Time: 85 Ma ago **Location:** Neuquen, Argentina **Habitat:** Bajo de la Carpa Formation **Size:** WS unknown; BL ~60cm (2ft); TL unknown **Features:** Neck long w/ large, round head. Wings very small. Pelvis broad. Legs long. Tarsus broad. Hallux oriented forward (not reversed) and apparently used in locomotion alongside other toes. Tail long & partially unfused. **Biology:** These appear to have been specialized flightless, slow-moving foragers. Much of the skeleton & skull unknown.

Wine-loving Gargantua Bird ***Gargantuavis philoinos***
Time: 70 Ma ago **Location:** Aude, France
Habitat: Marnes de la Maurine Formation. Ibero-Armorican island in the Tethys sea **Size:** WS unknown; BL ~2m (6.5ft); TL unknown **Features:** Known from hip & possible femur & neck bone. Pelvis very broad. Hip socket located near front of hip. **Biology:** Largest known Mesozoic avialans. Large, heavy bones indicate these were flightless ground birds. Broad pelvis unlike that in specialized runners, e.g. ostriches and terror birds, but characteristic of slow-moving moa & mihirungs. Probably more moa-like than ostrich-like in ecology, employing an herbivorous browsing lifestyle. A few features of internal anatomy may indicate a close relationship to *Patagopteryx*. The name refers to the fact that fossils were found while building a winery, which now produces a *Gargantuavis*-label pinot noir. Much of the anatomy is unknown, & so is restored similarly to ratites. This general body plan seems to have been converged upon several times among large flightless birds. Long neck & small head is suggested by anatomy of a possible neck vertebrae described in 2012.

Derived Euornitheans & Basal Carinatans

The *Carinatae* ("keeled" birds, referring to the prominent muscle attachment site on the breastbone) includes modern toothless birds (*Aves*), and their close, toothed relatives such as the ichthyornitheans, "fish birds". The latter were named, ironically, for the structure of their vertebrae, and not for the animals that almost certainly made up the bulk of their diet. Most ichthyornitheans are known from very fragmentary remains, and while some researchers have assigned all known remains to a single species, their geographic distribution and large temporal range, combined with variations in size over time, indicate that several species can potentially be recognized. Together with the aquatic hesperornitheans, the carinatans form the *Ornithurae*, an advanced group including the first known truly aquatic birds. The first known lineage to become specifically adapted to diving were the gansuids, currently known from the sole species *Gansus yumenensus*. *Gansus* may be primitive carinatans, or more basal euornitheans, though some informal analyses have found them to be true ichthyornitheans (Cau 2012).

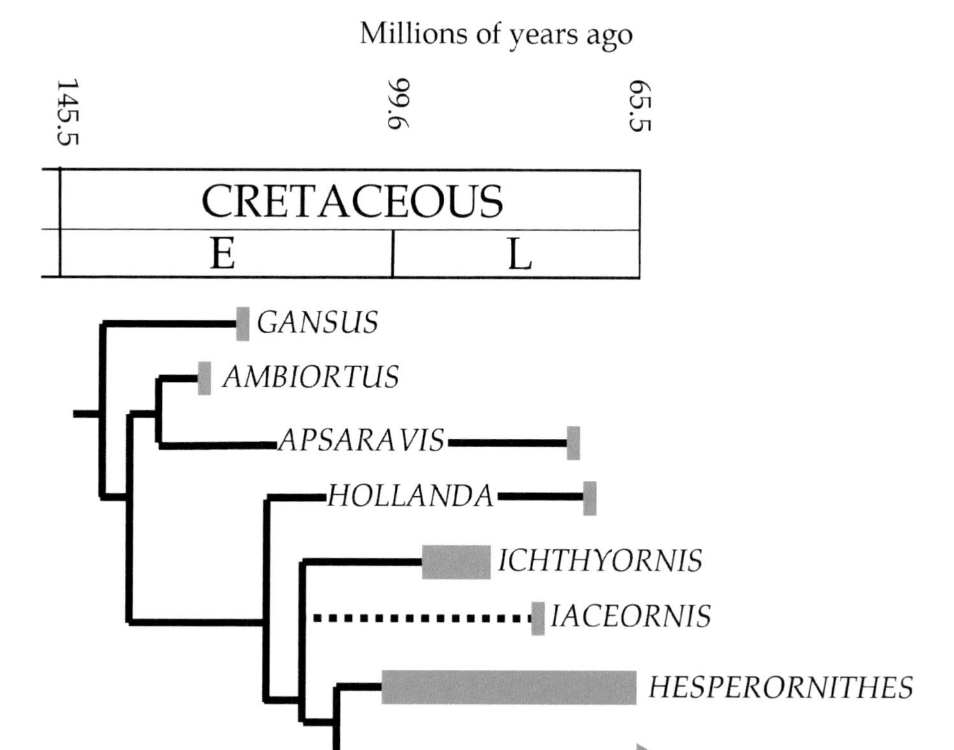

Above: Relationships of carinatans over time. Phylogeny approximated based on O'Connor & Zhou 2012.

Dement'yev's Vacillating Beginning ***Ambiortus dementjevi***
Time: 125 Ma ago **Location:** Bayankhongor, Mongolia **Habitat:** Andaikhudag Formation **Size:** WS unknown; BL ~25cm (in); TL unknown **Features:** Known from a fragmentary skeleton, distinguished by internal anatomy including slender shoulder blades & arrangement of ligament attachments on the humerus. **Biology:** Long, vaned primaries indicate flight capability. Generally very primitive, may be similar to yanornithiformes/ancestral carinatans.

Ukhaan Apsara Bird ***Apsaravis ukhaana***
Time: 75 Ma ago **Location:** Omnogovi, Mongolia **Habitat:** Ukhaa Tolgod, Djadochta Formation. **Size:** WS >30cm (1ft); BL ~18cm (7in); TL unknown **Features:** Head small but snout unknown. Wings moderately long (ulna & manus ~ half torso length) w/ advanced flight mechanism (causing hand to automatically extend during the flight stroke). Wing claws absent. Legs very long with long tarsus (over twice toe length). Toes relatively stout with large but weakly curved claws. **Biology:** Wide range of motion in toe joints usually found in species adapted to running or swimming. Desert ecosystem in which these lived would have been largely devoid of water during some parts of the year. Unlike most other Mesozoic euornitheans, which largely occupied marine or shoreline environments, *Apsaravis* appear to have become re-adapted to a ground-running, desert lifestyle. May have hunted lizards & other small vertebrates during dry seasons. Possibly close relatives of *Ambiortus* and *Palintropus*.

Shining Holland Bird ***Hollanda luceria***
Time: 72 Ma ago **Location:** Omnogovi, Mongolia **Habitat:** Hermin Tsav, Barun Goyot Formation. High desert. Dune fields prone to periodic flooding and arid scrublands. **Size:** WS unknown; BL ~50cm (in); TL unkown **Features:** Known only from hind limbs. Legs very long w/ long, slender & fused tarsus. Toes strong & robust. **Biology:** Anatomical comparisons using the legs & proportions of the toe bones indicate that these were specialized ground foragers, probably chasing fast-moving prey similar to modern roadrunners. Probably retained teeth in the rear of the jaws as in most other non-avian euornitheans.

Yumen Gansu *Gansus yumenensis*
Time: 120 Ma ago **Location:** Gansu, China **Habitat:** Xiagou Formation. Wide basin dominated by a system of large, tranquil freshwater lakes. **Size:** WS 40cm (1.3ft); BL ~24cm (9in); TL unknown **Features:** Wings large & powerful, relatively short and broad (humerus & ulna each ~ half torso length, manus slightly over half ulna length). Remiges asymmetrical, w/ primaries up to ~10cm long in large specimens, secondaries up to ~8cm. Small claw on alular digit. Minor digit fused to major digit. Breastbone strongly keeled. Legs long (femur & tarsus ~ equal in length, tibia nearly twice femur length), with long toes (equal to tarsus in length) & small, straight claws. Feet webbed up to bases of claws & covered in fine scales. **Biology:** Freshwater aquatic. Swimming &/or diving birds very common in their environment. Long webbed hindlimbs adapted to paddling along lake surface, possibly dipping for fish & invertebrates. Probably employed soaring flight while airborne. Studies of the poorly preserved feather traces in some specimens show that the plumage was primarily dark colored or black.

Two-headed Fish Bird *Ichthyornis anceps*
Time: 80 Ma ago **Location:** Kansas, Alabama, Texas, & California, USA; Manitoba, Canada **Habitat:** Smokey Hill Chalk Member, Niobrara Formation. Deep waters of the Western Interior Seaway **Size:** WS >60cm (2ft); BL ~30cm (1ft); TL unknown **Features:** Head large w/ long, straight jaws ending in beak supported by predentary bone in lower jaw. Numerous teeth set in middle of top & bottom jaws behind the toothless tips. Neck long. Wings long & robust (ulna slightly over ¾ torso length, humerus & manus nearly equal to ulna in length). Alular digit very short & immobile. Minor digit fused to major digit. All wing digits lack claws but probably retained keratin sheathes on at least alular & major digits as in avians. Breastbone strongly keeled. Legs very short (femur & tarsus equal in length, tibia slightly under half femur length, tibia slightly shorter than humerus) & very slender. Toes largely unknown. **Biology:** Marine. Extremely common along the Western Interior Seaway coast from Alberta to Alabama. Likely gull-like in ecology, employing shallow dives to catch fish beneath the surface. Feet may have been webbed. Small bill may have aided in catching or spearing fish. Wings probably long, pointed, & narrow allowing long periods of dynamic soaring. While usually considered small, known specimens represent a huge range of sizes, & the largest are as large or larger than the largest gulls. Size variation does not fall into distinct classes as would be expected if it represented several species, instead forming a continuum. Probably represents growth stages of single species. This suggests that even advanced non-avian carinatans grew relatively slowly compared to modern birds.

Marsh's Neglected Bird *Iaceornis marshi*
Time: 80 Ma ago **Location:** Kansas, USA **Habitat:** Smokey Hill Chalk Member, Niobrara Formation **Size:** WS unknown; BL ~25cm (10in); TL unknown **Features:** Similar in anatomy to *I. anceps*, differed internally in features of shoulder blade & wishbone. Breastbone strongly keeled. Legs much larger relative to wings than *I. anceps* (tibia longer than manus). At least major digit lacked bony claws. Minor digit completely fused to major digit. **Biology:** Though very similar to the contemporary *I. anceps*, probably closer to *Aves*. It is unknown whether or not these had teeth; teeth would be expected to be absent if they were avians, and their more advanced position suggests teeth would at least have been reduced.

Avians

Avians, "(modern) birds", represent the bird 'crown group', descendants of the most recent common ancestor of all birds alive today. This ancestor seems to have existed as early as 130 million years ago, if fragmentary remains such as those classified as *Gallornis* are indeed avians. All known avians lack teeth, and instead have extensive, often fully-fused beaks. It is likely that teeth were lost completely in a predecessor to the common ancestor of all modern bird groups; however, it is also possible that two or more avian lineages lost teeth independently of one another.

Numerous species of avians have been identified from the Mesozoic, mainly from the latest Cretaceous, but are based on remains too fragmentary to reconstruct their life appearance with any degree of confidence (see Appendix A). Due to this poor fossil record, there has historically been a debate regarding the timing of modern bird diversification. However, the likely presence of many modern bird "orders" in the Mesozoic suggests that avians had begun to diversify into many of their modern forms by the time of the K-Pg extinction event that ended the Mesozoic era.

Based on tentative interpretations of the fossil evidence, it is likely that members of the following modern bird groups existed before the K-Pg boundary: *Charadriiformes* (wading shorebirds like *Cimolopteryx*), *Anseriformes* (including *Vegavis*, above), *Galliformes* (including *Austinornis*), *Palaeognathae* (ratites and allies represented by *Limenavis* and possibly some early lithornithids), and *Pelecaniforms* (including *Torotix* and some cormorant-like birds). Grebes and rail-like birds (possibly ancestral to Cenozoic "terror birds") have also been reported from the latest Cretaceous. At least one species or lineage representing each of these groups survived the K-Pg extinction and diversified into all the remaining groups of modern birds in the Cenozoic Era. In the case of the anseriformes, at least two species or lineages survived the mass extinction (presbyornithids and the ancestors of modern ducks and geese represented by *Vegavis*).

In addition to these avian lineages, it is possible that one or more groups of non-avian birds also survived across the K-Pg boundary. One species from the Paleogene, *Qinornis paleocenica*, is known from fossils which show unfused bones in the tarsus, despite the interpretation of the specimen as an adult, a trait known only in non-avian birds. This species may represent a lineage of toothed carinate birds that survived for several million years in the Cenozoic.

Not included here is the diverse bird fauna of the Hornerstown Formation of New Jersey, which probably formed very shortly after the K-Pg boundary (as evidenced by an abundance of re-buried mosasaur remains). This ecosystem preserved an abundance of waterbirds including waders and representatives of most bird groups mentioned above, further evidence that the birds which survived the extinction were primarily waterbirds, and which later evolved into "higher land birds" to fill the vacant niches left by enantiornitheans and other theropods.

Above: Hypothetical restorations of select Mesozoic avians. Clockwise from top left: Austinornis lentus, Cimolopteryx rara, Limenavis patagonica, Torotix clemensi. *Not to scale.*

IAA Vega Bird ***Vegavis iaai***

Time: 65.5 Ma ago **Location:** Vega Island, Antarctica **Habitat:** Sandwich Bluff Member, Lopez de Bertodano Formation **Size:** WS unknown; BL ~25cm (10in); TL unknown **Features:** Known from a fragmentary skeleton severely damaged during preparation out of the rock matrix. Wing incompletely known. Legs relatively long, w/ tibia as long or longer than humerus. Tarsus about ¼ or ½ length of ulna (based on known radius length). **Biology:** The relatively long legs may indicate a wading lifestyle similar to the related Paleogene period presbyornithids, broad-billed, stilt-legged ducks more advanced than *V. iaai*. Numerous subtle characteristics of the skeleton show that this species was a primitive member of the duck, goose, & swan lineage (*Anatoidea*), possibly similar to modern basal anatoids like screamers (*Anhimidae*) & the magpie-geese (*Anseranatidae*), which diverged from the duck lineage before the evolution of the characteristic broad, flat bill of anatids & presbyornithids. The bill of *Vegavis* may therefore have been narrow & slightly hooked, like primitive anseriformes, or may have begun to broaden. This would have determined their exact method of feeding, but like all basal anseriformes, *V. iaai* probably foraged for soft plant material either in the water &/or in marshy shores & wetlands. Like screamers & magpie-geese, their feet may have been partially webbed.

Gregory's Polar Bird ***Polarornis gregorii***

Time: 65.5 Ma ago **Location:** Vega Island, Antarctica **Habitat:** Sandwich Bluff Member, Lopez de Bertodano Formation **Size:** WS unknown; BL ~80cm (2.6ft); TL unknown **Features:** Known from a partial skeleton including partial skull, vertebrae & limb bones. Toothless bill long, narrow & triangular. Head long & squat. Neck long (nearly equal to torso length when fully extended). Legs probably long, w/ very short femur (less than half skull length) & large tibia w/ strong muscle attachments. Tibia probably long, but most of it unknown. **Biology:** Advanced features of the skull & legs show that these birds were probably members of the modern loon lineage (*Gaviiformes*). The walls of the bones were relatively thick compared to modern loons, indicating a flightless or near-flightless, diving lifestyle. The wings may therefore have been small compared to modern loons. *Neogaeornis wetzeli*, another early loon relative from Chile, may be a senior synonym of this species or an unnamed, apparently flying species from Antarctica that has yet to be described.

Hesperornitheans

Given the fact that most primitive euornitheans occupied semi-aquatic shorebird niches, it is unsurprising that some groups would diversify into fully aquatic forms. The major lineage of aquatic Mesozoic birds was the *Hesperornithes*, or "westbirds". The clade *Hesperornithes* includes four major lineages of foot-propelled divers: enaliornithids, brodavids, baptornithids, and hesperornithids. The enaliornithids are a poorly known group of primitive divers that may represent an unnatural grouping including the ancestors of the hesperornithids and baptornithids. They, and the brodavids, probably retained some flight ability, and hesperornithean feathers found preserved in Albertan amber may belong to roosting enaliornithids or brodavids.

The hesperornithoids, comprising the most specialized families baptornithidae and hesperornithidae, include the most fully aquatic and marine-adapted birds that ever lived, some almost completely forsaking their ability to move about on land except to lay eggs. The leg and hip anatomy of hesperornithids is very similar to that of modern loons, and they probably employed similar foot-propelled diving techniques in pursuit of aquatic prey. The anatomy of the toes, particularly the toe joints, is very similar to modern lobe-toed birds like grebes, rather than web-footed birds. Specifically, the toes probably carried one single large lobe each, rather than multiple lobes as in coots. As in aerodynamic flight feathers, the lobes were probably asymmetrical with respect to the skeletal toes for hydrodynamic purposes. Like many specialized diving birds, many may have been colored with counter-shaded patterns of dark on top and light on bottom to camouflage them against surface and seafloor. The groove-set teeth of hesperornithoids are also seen in mosasaurs and may be a specialization for catching fast-moving prey in an open ocean environment. Additionally, their jaws appear to have been prokinetic, i.e. the upper jaw was able to move up and down relative to the base of the skull, as in many modern birds (Bhuler & al. 1987).

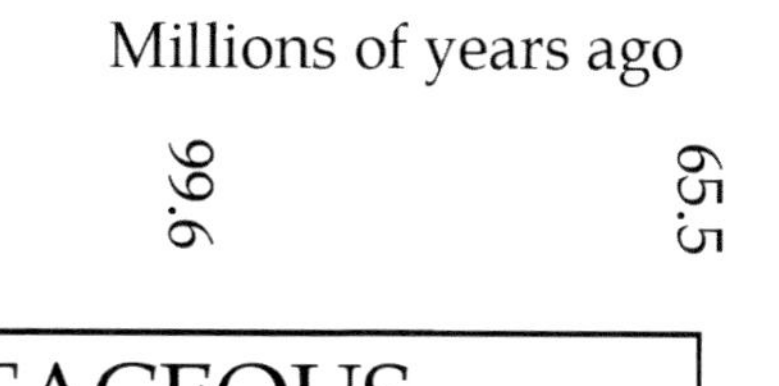

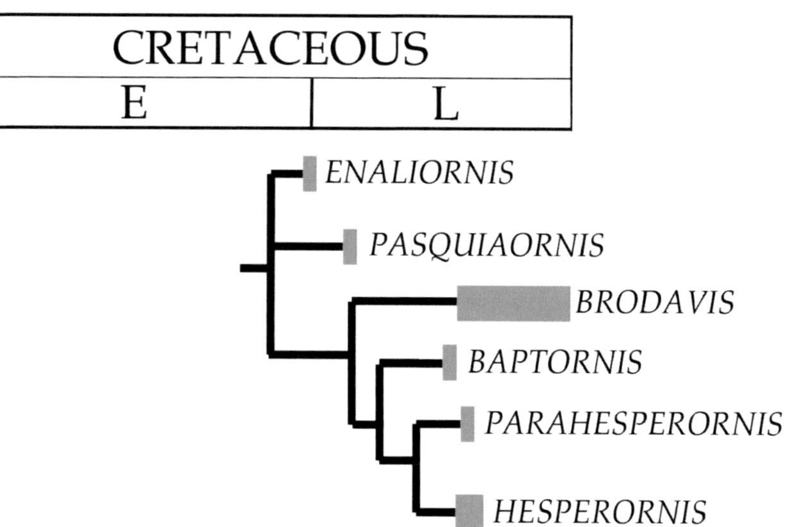

Above: Relationships of hesperornitheans over time. Phylogeny approximated based on Mortimer 2010.

Barrett's Seabird *Enaliornis barretti*
Time: 100 Ma ago **Location:** England, UK **Habitat:** Cambridge Greensand Formation. Shores and lagoons surrounding semi-tropical islands in a shallow sea. **Size:** WS unknown; BL ~55cm (1.8ft); TL unknown **Features:** Marine hesperornitheans known from a partial skeleton & a number of fragments. Head small, but most of the skull & snout unknown. Legs very long, w/ long, robust tibia, short & stocky femur (probably less than half tibia length), & short but robust tarsus shorter than the femur. Distinguished by their small size, primitive characteristics, & internal characteristics of the hind limbs & pelvis (lack of femoral neck, mid-sized antitrochantor, keel beneath the synsacrum). **Biology:** These small seabirds combined some features in common with more advanced hesperornitheans, while retaining some primitive features found in ichthyornitheans & other carinatans. They were therefore probably coastal divers or dippers which may have retained some flight ability, though the wing is not known.

Tanke's Pasquia Bird *Pasquiaornis tankei*
Time: 95 Ma ago **Location:** Saskatchewan, Canada **Habitat:** Ashville Formation, nearshore marine enviornment in the northern Western Interior Seaway **Size:** WS >36cm (1.2ft); BL ~70cm (2.3ft); TL unknown **Features:** Jaws long & toothed. Wings relatively long. Legs long & slender. **Biology:** The legs were positioned more directly under the body than in later hesperornitheans, possibly allowing a waddling gait on land. The toes could not twist during proportion as in lobed-toe birds, indicating normally webbed feet. The relatively long & robust wing bones suggest possible retention of flight ability, though the sternum is unknown. Probably surface swimmers (Sanchez 2012).

Varner's Brodkorb Bird *Brodavis varneri*
Time: 78 Ma ago **Location:** Kansas, USA **Habitat:** Sharon Springs Member, Pierre Shale Formation. Relatively shallow river deltas and estuaries opening into the Western Interior Seaway. **Size:** WS unknown; BL ~90cm (3ft); TL unknown **Features:** Marine hesperornitheans. Body long & barrel-shaped, w/ short torso & long hips. Legs very long, w/ tibia over half the length of the torso + hips. Tarsus short & stocky, less than ¼ tibia length. Neck long, probably equal in length to torso + hips. Wings unknown. **Biology:** The hip bones of these birds were not as advanced & highly fused as in related species. This, compared w/ the robust foot bones, may suggest a slightly different method of underwater propulsion compared w/ similar species inhabiting deeper saltwater environments off the coasts. The skeleton was overall more lightly built & not as solid as other hesperornitheans. This implies a more limited diving ability, restricting these birds to near-surface feeding. The lighter skeleton may also suggest some flight ability was retained.

Fallen Diving Bird ***Baptornis advenus***
Time: 80 Ma ago **Location:** Kansas, USA **Habitat:** Smokey Hill Chalk Member, Niobrara Formation. Deep waters of a warm inland sea dominated by ammonites and a diversity of small to gigantic fish. **Size:** WS 36cm (1.2ft); BL ~70cm (2.3ft); TL unknown **Features:** Marine. Head long & slender w/ very slim, toothed snout ending in a moderately long bill. Neck long & slender. Body long & barrel-shaped. Wings small & vestigial. Legs long, but w/ short tarsus & small feet. **Biology:** The smallest hesperornithean species in this environment, probably foraged for smaller prey and possibly in shallower water than relatives. Anatomy of the toe joints implies the feet were likely webbed, rather than lobed.

Alex's Near Western Bird ***Parahesperornis alexi***
Time: 80 Ma ago **Location:** Kansas, USA **Habitat:** Smokey Hill Chalk Member, Niobrara Formation **Size:** WS unknown; BL 1.1m (3.6ft); TL unknown
Features: Similar to, but more primitive than, *Hesperornis*. Toe anatomy consistant with lobed rather than webbed feet. **Biology:** Only known hesperornithean remains to preserve skin & feather impressions. Like grebes, tarsus was covered at the front w/ broad scutes near the feet (26 scutes in all). Closer to the body, the tarsus was covered in very long, plumulaceous feathers, the tips of which reached almost to the foot.

Thick-footed Western Bird ***Hesperornis crassipes***
Time: 80 Ma ago **Location:** Kansas, USA **Habitat:** Smokey Hill Chalk Member, Niobrara Formation **Size:** WS unknown; BL 1.1m (3.6ft); TL unknown **Features:** Marine, similar to *H. regalis* in anatomy. Differ in distinct tarsus w/ larger muscle attachmentssites at the ankle, & shallower attachment sites on the breastbone. **Biology:** Lived alongside related species, probably exploited different ecological niche. The differences in tarsal anatomy may indicate a different foot-propelled diving stroke.

Regal Western Bird ***Hesperornis regalis***
Time: 80-78 Ma ago **Location:** Kansas, USA **Habitat:** Smokey Hill Chalk Member, Niobrara Formation **Size:** WS ~46cm (1.5ft); BL 1.7m (5.5ft); TL unknown **Features:** Head small with long, thin jaws tipped w/ compound beak. Numerous teeth in jaws behind beak. Beak restricted to predentary on lower jaw, extensive in upper jaw, covering long premaxilla & continuing slightly above teeth. Beak slightly hooked. Lower jaw teeth set in grooves. Lower teeth locked into pits in premaxilla when jaws were closed. Body short & round w/ arched back. Wings very small, probably vestigial & hidden beneath body feathers. Upper legs attached to torso & likely affixed to body wall by soft tissue. Tarsi & feet emerged laterally from hip/tail area. Toes long, w/ fourth/inner toe longest. Toes probably lobed. Tail broad & flat, possibly somewhat beaver-like. **Biology:** Known from numerous remains incl. nearly complete skeletons. Like many large, deep-sea diving birds, they may have been counter-shaded for camouflage. While diving, necks were probably locked into place by ligaments in a tight curve with the head adjacent to the torso, to avoid torsion during quick underwater turning. Given the degree to which the hind limbs were permanently splayed & incorporated into the streamlined body wall similar to loons, these were probably extremely awkward on land, unable to walk, instead pushing along on the belly w/ a dragging motion similar to seals (Martin & al. 2012).

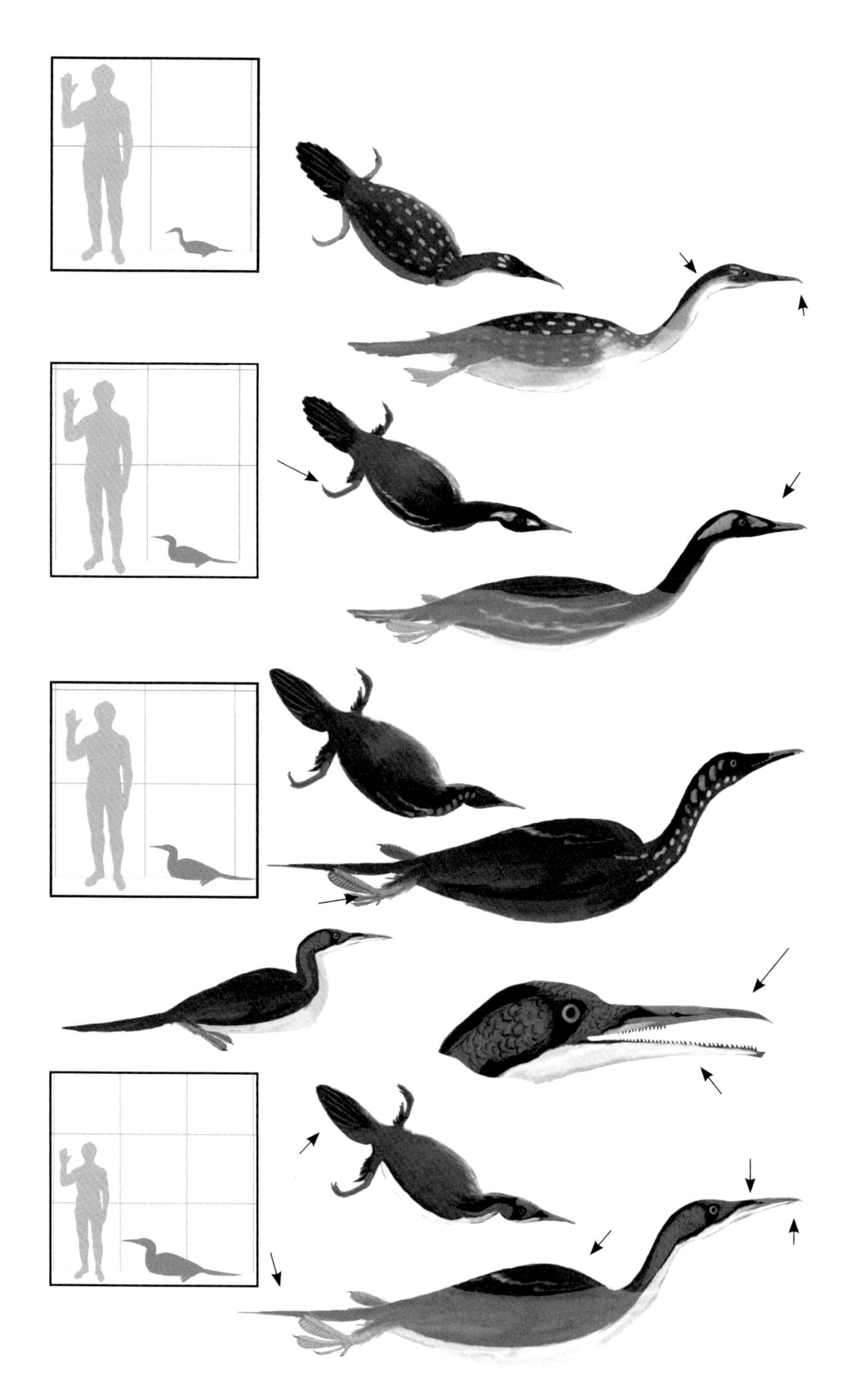

Appendix A: Excluded Species

Unfortunately, many fossil bird species have been named based on extremely fragmentary remains, or remains which cannot be restored because their classification is too uncertain to base restorations on their relatives. These species are briefly discussed below.

Indeterminate or Possible Birds

- *Cerebavis cenomanica* - known only from the cast of a brain case.
- *Cretaaviculus sarysuensis* - known from a single tiny but asymmetrical Stage V body feather, and possibly an isolated ornithothoracine claw from the same formation.
- *Elopteryx nopcsai* - known from a partial femur. May come from a bird or a more basal maniraptoran.
- *Hulsanpes perlei* - known from the partial hindlimbs of a hatchling specimen, possibly an deinonychosaurian or euornithean.
- *Ilerdopteryx viai* - known from isolated Stage IV feathers.
- *Koparion douglassi* - known only from teeth, may be a troodontid or compsognathid.
- *Kuszholia mengi* - known only from partial hip bones, an indeterminate bird, possibly a caenagnathiform.
- *Pneumatoraptor fodori* - known from a partial shoulder girdle with evidence of avian-type air sacs. *Bauxitornis mindszentyae*, named for a tarsometatarsus from the same location, may be a synonym if it is a *Balaur*-like basal eumaniraptoran and not an enantiornithean as suggested by Cau (online 2010b).
- *Praeornis sharovi* - known from isolated feathers possibly representing Prum's Stage IIIa.
- *Variraptor mechinorum* - known from a fragmentary skeleton, possibly deinonychosaurian.
- *Vorona berivotrensis* - known from a partial hind limb. May be enantiornitheans or primitive euornitheans.
- *Wyleyia valdensis* - known only from a humerus, it is some kind of non-euornithean bird.

Caenagnathiformes

- *Hagryphus giganteus* - large caenagnathid known only from wing bones.
- *Machairasaurus leptonychus* - oviraptorid known only from from wing bones.
- *Ojoraptorsaurus boerei* - caenagnathid known only from partial hip bones.

Deinonychosaurs

- "Coelurus" *gracilis* - Known from fragmentary remains including claws and teeth. Likely eudromaeosaurian.
- *Dromaeosauroides bornholmensis* - earliest eudromaeosaurian, known from a single tooth.
- "Euronychodon" *asiaticus* - known only from teeth, probably from a troodontid.
- "Euronychodon" *portucalensis* - known only from teeth, probably from a troodontid.
- *Itemirus medullaris* - Known initially from an isolated braincase similar to that of *Velociraptor*. Additional skeletons have yet to be described but are reported to approach *Utahraptor* in size (Sues & Averianov, 2004).
- *Ornithodesmus cluniculus* - known from a synsacrum.
- "Paronychodon" *caperatus* - known only from teeth, probably troodontid.
- *Paronychodon lacustris*- known only from teeth, probably troodontid.
- *Pectinodon bakkeri* - known only from teeth which are indistinguishable from *Troodon formosus* but much later in time.
- *Pyroraptor olympius* - known from teeth and fragmentary foot, wing bones, and vertebrae.
- *Richardoestesia gilmorei* - teeth, possibly from a species of microraptorians.
- *Richardoestesia isosceles* - teeth, possibly from a species of microraptorians.
- "Saurornithoides" *asiamericanus* - Known from teeth that differ from *Troodon formosus* in having flatter sides and larger serrations with narrow bases.
- *Saurornitholestes robustus* - known from a skull fragment which cannot be relaibly distinguished from other ornithodesmid remains.
- *Tochisaurus nemegtensis* - known from metatarsals, probably troodontid.
- *Urbacodon itemirensis* - teeth and skeletal fragments. Probably troodontid.
- *Zapsalis abradens* - teeth, possibly from small or juvenile dromaeosaurines.

Enantiornitheans

- *Abavornis bonaparti* - known from a partial coracoid.
- *Alexornis antecedens* - known from a partial wing and other fragments.
- *Avisaurus gloriae* - known from a tarsometatarsus.
- *Catenoleimus anachoretus* - known from a partial coracoid.
- "Cathayornis" *caudatus* - dubious species probably synonymous with a contemporary.
- *Enantiornis leali* - known only from wing bones, it is probably synonymous with *Lectavis bretincola*, known only from leg bones.
- *Dalingheornis liweii* - known from a juvenile specimen and probably synonymous with a contemporary. The supposed long, unfused tail is a juvenile characteristic (as in *Zhongornis*) and the reported zygodactyl foot may an artifact of preservation.
- *Elsornis keni* - known from wing bones, the proportions indicate these were the only known flightless enantiornitheans.
- *Explorornis nessovi* - known from a partial coracoid.
- *Explorornis walkeri* - known from a partial coracoid.
- *Flexomornis howei* - known from a partial wing and other fragments.
- *Gurilynia nessovi* - known from a partial humerus and coracoid.
- "Ichthyornis" *minusculus* - known from a single verteba.

- *Incolornis martini* - known from a partial coracoid.
- *Incolornis silvae* - known from a partial coracoid.
- *Kizylkumavis cretacea* - known from a partial humerus.
- *Lenesornis maltshevskyi* - known from a partial sacrum.
- *Liaoxiornis delicatus* - known from hatchling specimens, probably synonymous with a contemporary.
- *Martinavis cruzyensis* - known from an isolated humerus.
- *Martinavis minor* - known from a partial humerus.
- *Martinavis saltariensis* - known from a humerus, probably synonymous with another Lecho formation species.
- *Martinavis vincei* - known from humeri, probably synonymous with another Lecho formation species.
- *Martinavis whetstonei* - known from a partial humerus.
- *Nanantius eos* - known from a partial tibiotarsus and isolated vertebra.
- *Noguerornis gonzalezi* - known from a partial wing w/ feather impressions and skeletal fragments.
- *Platanavis nana* - known from a partial sacrum.
- *Sazavis prisca* - known from a partial tibiotarsus.
- *Xiangornis shenmi* - known from a partial wing and shoulder girdle.

Euornitheans

- *Alamitornis minutus* - possible patagopterygiforms known from partial humeri.
- *Apatornis celar* - known from a partial synsacrum, possibly avian.
- *Asiahesperornis bazhanovi* - hesperornitheans known from fragmentary specimens.
- *Austinornis lentus* - a possible galliform known from a partial metatarsus.
- *Brodavis americanus* - hesperornitheans known from the metatarsus.
- *Brodavis baileyi* - hesperornitheans known from the metatarsus.
- *Brodavis mongoliensis* - hesperornitheans known from the metatarsus.
- *Canadaga arctica* - giant hesperornitheans known from vertebra and a partial femur.
- *Ceramornis major* - avians known from a partial coracoid.
- *Cimolopteryx maxima* - charadriiform known from a partial coracoid.
- *Cimolopteryx minima* - charadriiform known from a partial coracoid.
- *Cimolopteryx petra* - charadriiform known from a partial coracoid.
- *Cimolopteryx rara* - charadriiform known from a partial carpometacarpus.
- *Chaoyangia beishanensis*- known from a partial hind limb and pelvis, probably similar to *Zhongjianornis*.
- *Enaliornis sedgewicki* - small hesperornitheans known from hind limbs.
- *Enaliornis seeleyi* - small hesperornitheans known from hind limbs.
- *Eurolimnornis corneti* - known from a partial wing.
- *Gallornis straeleni* - possible avians known from partial hind limb.
- *Graculavis augustus* - a possible charadriiform.
- *Guildavis tener* - known from a partial synsacrum.
- *Hesperornis altus* - hesperornithids known from a partial tarsometatarsus.
- *Hesperornis bairdi* - hesperornithids known from partial hips and tarsals.
- *Hesperornis chowi* - hesperornithids known from a partial tarsometatarsus.
- *Hesperornis gracilis* - hesperornithids known from a partial hindlimb.

- *Hesperornis macdonaldi* - very small hesperornithids known from an isolated femur.
- *Hesperornis mengeli* - hesperornithids known from a tibiotarsus.
- *Hesperornis montanus* - hesperornithids known from a single vertebra.
- *Hesperornis rossicus* - hesperornithids known from a partial hindlimb.
- *Horezmavis eocretacea* - known from a partial tarsometatarsus.
- *Judinornis nogontsavensis* - long-bodied hesperornitheans known from a partial vertebra.
- *Limenavis pategonica* - known from a partial wing, possible lithornithids.
- *Lonchodytes estesi* - a procellariform.
- "Lonchodytes" *pterygius* - a possible charadriiform.
- *Neogaeornis wetzeli* - a possible gaviiform known from the tarsus.
- *Palaeocursornis corneti* - known from a partial femur.
- "Palaeotringa" *vetus* - a possible presbyornithid anseriform known from a partial tibia.
- *Palintropus retusus* - a possible ambiortiform known from a partial coracoid.
- *Parascaniornis stenseoei* - hesperornitheans known from the tibiotarsus and vertebra.
- *Pasquiaornis hardiei* - small hesperornitheans known from partial hind limbs.
- *Piksi barbarulna* - known from a partial wing, possibly a basal euornithean.
- *Potamornis skutchi* - primitive hesperornitheans known from partial hind limbs.
- *Telmatornis priscus* - possibly similar to a great crested grebe.
- *Teviornis gobiensis* - known from a partial wing, possible presbyornithids.
- *Torotix clemensi* - known from a partial humerus, a possible pelecaniform.
- *Volgavis marina* - possible charadriiform known from a beak.
- *Zhyraornis kashkarovi* - known from a partial synsacrum.
- *Zhyraornis logunovi* - known from a partial synsacrum.

Appendix B: Clade Definitions

The exact relationships of prehistoric birds are contentious and frequently change with new phylogenetic hypotheses, so it is especially important to define the taxonomic terms we use to discuss them. The operational definitions for the groups used in this book are given below.

Because the International Code of Phylogenetic Nomenclature (ICPN, or "PhyloCode") has not gone into effect as of this writing, the oldest available phylogenetic definition for each group is used. Notation follows the recommendations of ICPN Article 9.3.1. Note that some of the definitions are redundant: Definitions for junior synonyms are given for completeness if they are well-known groups. In the spirit of ICPN Recommendation 11A, any new definitions will include taxa included by the nominal authors as specifiers where practical. New definitions are proposed for groups which have name priority over equivalent groups or for useful taxa that have not previously been defined.

Aviremigia Gauthier & de Queiroz 2001 [Gauthier & de Queiroz 2001]
> Remiges and rectrices (enlarged, stiff-shafted, closed-vaned with barbules bearing hooked distal pennulae), pennaceous feathers arising from the distal forelimbs and tail
*Note: Presumably anchored on these characters in *Passer domesticus*

Chuniaoae Ji & al. 1998 [Ji & al. 1998]
< *Caudipteryx* & *Avialae*

Caenagnathiformes Sternberg 1940 [converted clade name]
> *Caenagnathus collinsi* ~ *Passer domesticus*

Oviraptorosauria Barsbold 1976 [Currie & Padian 1997]
> *Oviraptor* ~ "Birds"
*Note: "Birds" presumably crown Aves

Caenagnathoidea Sternberg 1940 [Sereno 1999]
< *Caenagnathus & Oviraptor*

Caenagnathidae Sternberg 1940 [Sues 1997]

< *Caenagnathus pergracilis & Chirostenotes elegans & Elmisaurus rarus & Caenagnathasia martinsoni &* BHM 2033

Oviraptoridae Barsbold 1976 [Sereno 1998]
> *Oviraptor ~ Caenagnathus*

Oviraptorinae Barsbold 1976b [Osmólska & al. 2004]
< *Oviraptor philoceratops & Citipati osmolskae*

Caenagnathinae Paul 1988 [converted clade name]
> *Caenagnathus collinsi ~ Oviraptor philoceratops, Avimimus portentosus*

Avimimidae Kurzanov 1981 [converted clade name]
> *Avimimus portentosus ~ Oviraptor philoceratops, Elmisaurus rarus, Caenagnathus collinsi*

Eumaniraptora Padian & al. 1997 [Padian & al. 1999]
< *Deinonychus & Neornithes*

Ornithes new clade name [new definition]
< *Archaeopteryx lithographica & Passer domesticus*

Saurornithes Nicholson 1878 [converted clade name]
> *Archaeopteryx lithographica ~ Passer domesticus*

Deinonychosauria Colbert & Russell 1969 [Padian 1997]
> *Deinonychus* ~ "Birds"
*Note: "Birds" presumably meant crown *Aves*

Archaeopterygidae Huxley 1871 [Xu & al. 2011]
> *Archaeopteryx lithographica ~ Passer domesticus, Dromaeosaurus albertensis*

Ornithodesmiformes new clade name [new definition]
< *Ornithodesmus cluniculus & Dromaeosaurus albertensis & Troodon formosus, Archaeopteryx lithographica*

Troodontidae Gilmore 1924 [Varricchio 1997]
> *Troodon formosus & Saurornithoides mongoliensis & Borogovia gracilicrus & Sinornithoides youngi ~ Ornithomimus velox, Oviraptor philoceratops*, "other well-defined groups"
*Note: "other well defined groups" presumably include eudromaeosaurs and crown avians

Troodontinae new clade name [new definition]
< *Troodon formosus & Saurornithoides mongoliensis*

Ornithodesmidae Hooley 1913 [converted clade name]
> *Ornithodesmus cluniculus ~ Archaeopteryx lithographica, Passer domesticus, Parony-*

chodon lacustris, Pterodactylus antiquus

Dromaeosauridae Russell 1969 [Sereno 1998]
> *Velociraptor ~ Troodon*

Microraptoria Senter & al. 2004 [Senter & al. 2004]
> *Microraptor ~ Velociraptor, Dromaeosaurus*

Microraptorinae Senter & al. 2004 [converted clade name]
> *Microraptor zhaoianus ~ Sinornithosaurus millenii, Dromaeosaurus albertensis*

Unenlagiinae Makovicky Apesteguia & Agnolin 2005 [Makovicky Apesteguia & Agnolin 2005]
> *Unenlagia comahuensis ~ Velociraptor mongoliensis*

Eudromaeosauria Longrich & Currie 2009 [Longrich & Currie 2009]
< *Saurornitholestes langstoni & Velociraptor mongoliensis & Deinonychus antirrhopus & Dromaeosaurus albertensis*

Itemiridae Kurzanov 1976 [converted clade name]
> *Itemirus medullaris ~ Dromaeosaurus albertensis, Stenonychosaurus inequalis, Tyrannosaurus rex*
*Note: *Stenonychosaurus inequalis* is a junior synonym of *Troodon formosus*

Velociraptorinae Barsbold 1983 [Sereno 1998]
> *Velociraptor ~ Dromaeosaurus*

Dromaeosaurinae Matthew & Brown 1922 [Sereno 1998]
> *Dromaeosaurus ~ Velociraptor*

Ornithurae Haekel 1866 [Gauthier 1986]
> *Aves ~ Archaeopteryx*

Avialae Gauthier 1986 [Gauthier 1986]
> *Ornithurae ~ Deinonychosauria*

Scansoriopterygidae Czerkas & Yuan 2002 [Zhang & al. 2008]
< *Epidexipteryx & Epidendrosaurus*
*Note: *Epidendrosaurus* is a junior synonym of *Scansoriopteryx*

Avebrevicauda Paul 2002 [Paul 2002]
> free caudals reduced to ten or fewer (*Neornithes*)
*Note: *Neornithes* is a junior synonym of *Aves*

Omnivoropterygiformes Czerkas & Ji 2002 [converted clade name]
> *Omnivoropteryx sinousaorum ~ Passer domesticus*
*Note: *Omnivoropteryx* is a junior synonym of *Sapeornis*, however it is used as a specifier because its eponymous "order" level name is older than that of *Sapeornis*.

Pygostylia Chatterjee 1997 [Chiappe 2001]
< *Confuciusornithidae & Neornithes*

Confuciusornithiformes Hou & al. 1995 [converted clade name]
> *Confuciusornis sanctus ~ Passer domesticus, Enantiornis leali*

Confuciusornithidae Hou & al. 1995 [Chiappe & al. 1999]
< *Confuciusornis sanctus & Changchengornis hengdaoziensis*

Ornithothoraces Chiappe & Calvo 1994 [Chiappe 1995]
< *Iberomesornis & Neornithes*

Ornithuromorpha Chiappe & al. 1999 [Chiappe 2001]
< *Vorona & Patagopteryx &* "Ornithurae"
*Note: "Ornithurae" meaning *Hesperornis & Aves*

Enantiornithes Walker 1981 [Sereno 1998]
> *Sinornis ~ Neornithes*

Iberomesornithiformes Sanz & Bonaparte 1992 [converted clade name]
> *Iberomesornis romeralii ~ Cathayornis yandica, Gobipteryx minuta, Enantiornis leali*

Euenantiornithes Chiappe 2002 [Chiappe 2002]
> *Sinornis ~ Iberomesornis*

Alexornithiformes Brodkorb 1976 [converted clade name]
> *Alexornis antecedens ~ Coracias garrulus, Picus viridis, Gobipteryx minuta*

Eoenantiornithiformes Hou & al. 1999 [converted clade name]
> *Eoenantiornis buhleri ~ Cathayornis yandica, Iberomesornis romeralii, Enantiornis leali*

Longipterygiformes Zhang & al. 2001 [converted clade name]
> *Longipteryx chaoyangensis ~ Cathayornis yandica, Iberomesornis romeralii, Enantiornis leali*

Eoenantiornithidae Hou & al. 1999 [converted clade name]
> *Eoenantiornis buhleri ~ Longipteryx chaoyangensis, Cathayornis yandica, Enantiornis leali*

Longipterygidae Zhang & al. 2001 [O'Connor & al. 2009]
< *Longipteryx chaoyangensis & Longirostravis hani*

Cathayornithiformes Zhou Jin & Zhang 1992 [converted clade name]
> *Cathayornis yandica ~ Iberomesornis romeralii, Longipteryx chaoyangensis, Gobipteryx minuta, Enantiornis leali*

Avisauroidea new clade name [new definition]
> *Avisaurus archibaldi* ~ *Longipteryx chaoyangensis, Sinornis santensis, Gobipteryx minuta*

Avisauridae Paul & Brett-Surman 1985 [Chiappe 1993]
< *Avisaurus archibaldi & Neuquenornis volans*

Enantiornithiformes Walker 1981 [converted clade name]
> *Enantiornis leali* ~ *Gobipteryx minuta*

Euornithes Cope 1889 [Sereno 1998]
> *Neornithes* ~ *Sinornis*

Patagopterygiformes Alvarenga & Bonaparte 1992 [converted clade name]
> *Patagopteryx deferrariisi* ~ *Passer domesticus*

Chaoyangiformes Hou 1997 [converted clade name]
> *Chaoyangia beishanensis* ~ *Passer domesticus*

Songlingornithidae Hou 1997 [converted clade name]
> *Songlingornis linghensis* ~ *Chaoyangia beishanensis, Passer domesticus*
*Note: *Songlingornis linghensis* may be a junior synonym of *Chaoyangia beishanensis*. However, the intent of the group *Songlingornithidae* was to separate those two species. That intent is preserved in this definition, which causes the clade *Songlingornithidae* itself to become a junior synonym of *C. beishanensis* if it is synonymous with *S. linghensis.*

Yanornithiformes Zhou & Zhang 2001 [converted clade name]
> *Yanornis martini* ~ *Passer domesticus*

Ambiortiformes Kurochkin 1982 [converted clade name]
> *Ambiortus dementjevi* ~ *Passer domesticus*

Apsaraviformes Livezey & Zusi 2007 [converted clade name]
> *Apsaravis ukhanna* ~ *Passer domesticus*

Palintropiformes Longrich Tokaryk & Field 2011 [Longrich Tokaryk & Field 2011]
> *Palintropus* ~ *Passer, Hesperornis, Ichthyornis*

Odontoclae Marsh 1875 [converted clade name]
> Teeth set in grooves in *Hesperornis regalis*

Odontornithes Marsh 1873 [converted clade name]
< *Ichthyornis anceps & Hesperornis regalis, Passer domesticus*

Hesperornithes Sharpe 1899 [Clarke 2004]
> *Hesperornis regalis* ~ *Aves*

Hesperornithiformes Sharpe 1899 [converted clade name]
< *Hesperornis regalis & Enaliornis barretti*

Enaliornithidae Fürbringer 1888 [converted clade name]
> *Enaliornis barretti ~ Hesperornis regalis*

Brodavidae Martin & al. 2012 [converted clade name]
> *Brodavis americanus ~ Hesperornis regalis*

Hesperornithoidea Shufeldt 1903 [converted clade name]
< *Hesperornis regalis & Baptornis advenus*

Baptornithidae AOU 1910 [converted clade name]
> *Baptornis advenus ~ Hesperornis regalis*

Hesperornithidae Marsh 1872 [Clarke 2004]
> *Hesperornis regalis ~ Baptornis advenus*

Carinatae Merrem 1813 [Carcraft 1986]
> *Neornithes ~ Hesperornis*

Gansuiformes Hou & Liu 1984 [converted clade name]
> *Gansus yumenensis ~ Passer domesticus, Hesperornis regalis, Ichthyornis anceps, Enantiornis leali*

Ichthyornithes Marsh 1873b [Clarke 2004]
> YPM 1450 ~ Aves
*Note: YPM 1450 is the holotype of *Ichthyornis dispar*

Ichthyornithiformes Fürbringer 1888 [converted clade name]
> *Ichthyornis anceps ~ Hesperornis regalis, Gansus yumenensis, Passer domesticus*

Aves Linnaeus 1758 [Gauthier 1986]
< *Ratitae & Tinamidae & Neognathae*

Neornithes Gadow 1892 [Sereno 1998]
< *Struthio & Passer*

Galloanserae Sibley & al. 1988 [Gauthier & de Queiroz 2001]
< *Gallus gallus & Anser anser*

Anseriformes Wagler 1830 [converted clade name]
> crown *Anser anser ~ Gallus gallus, Passer domesticus*
*Note: crown extant as of the year 1830

Anatoidea Leach 1820 [converted clade name]
> *Anser anser ~ Anseranas semipalmata*

Gaviiformes Wetmore & Miller 1926 [converted clade name]
> crown *Gavia immer* ~ *Podiceps cristatus, Passer domesticus*
*Note: crown extant as of the year 1926

Charadriiformes Huxley 1867 [converted clade name]
> crown *Charadrius hiaticula* ~ *Passer domesticus*
*Note: crown extant as of the year 1867

Appendix C: Evolutionary Linnaean Classification

Several attempts have been made in recent years to construct traditional or Linnaean classifications of Mesozoic birds. However, most of these attempts have been flawed in some way. For example, Livezey & Zusi (2007) presented a classification of all birds which included only a few Mesozoic taxa. These were united in several unorthodox, often paraphyletic groups which contradicted most contemporary phylogenies (e.g. uniting *Archaeopteryx* and *Confuciusornis* in a single group to the exclusion of other birds, including Rahonavis as an euenantiornithean etc.). While Linnaean classifications are clearly not as useful as phylogenies for elucidating the interrelationships of taxa, there are situations where they may be useful for communication. For that reason they should be made as comprehensive and as phylogenetically rigorous as possible within the confines of the Linnaean structure, and for these reasons Livezey & Zusi's classification is considered inadequate.

Another Linnaean classification was presented by Benton (2004). While more phylogenetically rigorous than that of Livezey & Zusi (2007), it suffered from oversimplification due to few included taxa. Presented here is a more comprehensive and rigorous Linnaean taxonomy of all known Mesozoic bird taxa (Subfamily rank and above). The goal of this taxonomy is not to be used in preference to phylogenetic nomenclature, but to be available as an alternative to published Linnaean classifications on the occasion that the use of Linnaean ranks is necessary or preferred by the author. The names and placement of taxa listed here are informed by their phylogenetic definitions in Appendix B.

- Class Aves Linnaeus 1758
  - Order Caenagnathiformes Sternberg 1940
    - Family Caudipteridae Zhou & Wang 2000
    - Family Avimimidae Kurzanov, 1981
    - Superfamily Caenagnathoidea Sternberg 1940
      - Family Caenagnathidae Sternberg 1940

Subfamily Caenagnathinae Sternberg 1940
Subfamily Elmisaurinae Osmolska 1981
Family Oviraptoridae Osborn 1924
Subfamily Oviraptorinae Osborn 1924
Subfamily "Ingeniinae" Barsbold 1981 (preocc.)
Order Deinonychosauria Colbert & Russell 1969
Family Troodontidae Gilmore 1924
Subfamily Jinfengopteryginae Turner & al. 2012
Subfamily Troodontinae Gilmore 1924
Subfamily Saurornithoidinae Barsbold 1974
Family Ornithodesmidae Hooley 1913
Subfamily Microraptorinae Longrich & Currie 2009
Subfamily Unenlagiinae Makovicky & al. 2005
Subfamily Saurornitholestinae Longrich & Currie 2009
Subfamily Itemirinae Kurzanov 1976
Subfamily Dromaeosaurinae Matthew & Brown 1924
Family Scansoriopterygidae Czerkas & Yuan 2002
Family Archaeopterygidae Huxley 1871
Family Jeholornithidae Zhou & Zhang 2006
Family Yandangornithidae Cai & Zhou 1999
Family Omnivoropterygidae Czerkas & Ji 2002
Order Confuciusornithiformes Hou & al. 1995
Family Confuciusornithidae Hou & al. 1995
Subclass Enantiornithes Walker 1981
Order Iberomesornithiformes Sanz & Bonaparte 1992
Family Iberomesornithidae Sanz & Bonaparte 1992
Family Liaoningornithidae Hou 1996
Family Protopterygidae Zhang & Zhou 2006
Family Alexornithidae Brodkorb 1976
Family Gobipterygidae Elzanowski 1974
Order Eoenantiornithiformes Hou & al 1999
Family Eoenantiornithidae Hou & al 1999
Family Longipterygidae Zhang & al. 2001
Order Cathayornithiformes Zhou & al. 1992
Family Cathayornithidae Zhou & al. 1992
Superfamily Avisauroidea Paul & Brett-Surman 1985
Family Mystiornithidae Kurochkin & al. 2011
Family Concornithidae Kurochkin 1996
Family Avisauridae Paul & Brett-Surman 1985
Order Enantiornithiformes Martin 1983
Family Enantiornithidae Nessov 1984
Order Chaoyangiformes Hou 1997
Family Patagopterygidae Alvarenga & Bonaparte 1992
Family Hongshanornithidae O'Connor & al. 2009
Family Songlingornithidae Hou 1997
Family Eurolimnornithidae Kessler & Jurcsak 1986
Order Palaeocursornithiformes Kessler & Jurcsak 1986
Family Palaeocursornithidae Kessler & Jurcsak 1988

Family Gansuidae Hou & Liu 1984
Family Ambiortidae Kuochkin 1982
Subclass Hesperornithes Marsh 1875
    Family Enaliornithidae Fürbringer 1888
    Family Brodavidae Martin & al. 2012
    Superfamily Hesperornithoidea Shufeldt 1903
        Family Baptornithidae AOU 1910
        Family Hesperornithidae Marsh 1872
            Subfamily Asiahesperornithinae Nessov & Prizemlin 1991
            Subfamily Hesperornithinae Marsh 1872
Subclass Ichthyornithes Marsh 1873b
    Family Ichthyornithidae Marsh 1873a
Subclass Neornithes Gadow 1893
    Superorder Palaeognathae Pycraft 1900
    Superorder Neognathae Pycraft 1900
        Order Anseriformes Wagler 1831
            Superfamily Anatoidea Vigors 1825
                Family Presbyornithidae Wetmore 1926
        Order Galliformes Temminck 1820
        Order Gaviiformes Wetmore & Miller 1926
        Order Pelecaniformes Sharpe 1891
            Family Torotigidae Brodkorb1963
        Order Charadriiformes Huxley 1867
            Family Cimolopterygidae Brodkorb 1963
        Order Cariamiformes Fürbringer 1888

Glossary

- *Alula*: Vaned, pennaceous feathers extending from the alular digit, also "bastard wing"
- *Alular digit*: The first digit finger of the hand
- *Arboreal*: Tree-dwelling
- *Basal*: A species or group positioned near the base of its parent clade's family tree
- *Barb*: Thin filament branching from the rachis and forming the vane of a feather
- *Barbule*: Small filaments branching from the barbs, adhering to adjacent barbs via hooklets
- *Bastard wing*: See *alula*
- *Clade*: A natural group, consisting of two specified sub-groups, their concestor, and all of its descendants
- *Closed-vaned feather*: See *vaned feather*
- *Concestor*: The most recent common ancestor of two given species or clades
- *Contour feather*: Vaned feathers covering the body
- *Covert*: Small vaned, pennaceous feathers covering the bases of remiges or rectrices
- *Crown*: Feathers covering the top of the skull, especially when forming a raised crest
- *Derived*: A species or group positioned far from the base of its parent clade's family tree
- *Digit*: "Finger" of the hand (manual digit) or "toe" of the foot (pedal digit)
- *Down*: Feather with a short or thin rachis and soft barbs lacking barbules
- *Fan-tail*: Rectrices arranged in a fan attaching to a pygostyle with rectrical bulb
- *Frond-tail*: Rectrices arranged in pairs along the length of a tail with discrete vertebrae
- *Hallux*: Fouth digit of the foot, usually reversed (backward-pointed and opposable) in perching species
- *Hindwing*: Wing-like structure formed by vaned feathers attached in a planar arrangement to the tarsus
- *Hooklet*: Microscopic hook-shaped filament holding barbs and barbules together in the vane of a feather
- *Humerus*: Bone of the upper arm, to which tertial feather ligaments anchor
- *Major digit*: The second (usually largest) finger of the hand
- *Manus*: Hand including the metacarpals and digits
- *Minor digit*: The third finger of the hand, often reduced and/or fused to the major digit
- *Open-vaned feather*: Pennaceous feathers lacking barbules, but in which the barbs are large and relatively stiff, forming a loosely planar surface
- *Propatagium*: Skin and ligaments connecting the wrist to the shoulder in a wing
- *Pygostyle*: Fused tail vertebrae, anchoring the rectrical bulb and rectrices

- *Quill*: See *rachis*
- *Rachis* (pl. *rachides*): Central shaft or "quill" of a feather
- *Radius*: Leading-edge bone of the forearm/wing
- *Rectrical bulb*: Muscles attached to the pygostyle which control the folding of the rectrices
- *Rectrix* (pl. *rectrices*): Main vaned, pennaceous feathers of the tail
- *Remix* (pl. *remiges*): Main vaned, pennaceous feathers of the wing (anchored to the ulna)
- *Ribbon-tail*: Rectrices arranged in pairs at the tip of a short tail with fused vertebrae, usually with vanes restricted to the tips
- *Scapular*: Long, vaned feathers attached to the shoulder which paritally cover the wing when folded
- *Tarsus*: Lowest portion of the leg, formed from the metatarsal and tarsal bones
- *Tertial*: Feathers partially filling in the gap between the remiges of the wing and the contour feathers of the body (anchored to the humerus)
- *Ulna*: Trailing bone in the forelimb/wing, to which secondary feather ligaments attach; usually bowed in flying species
- *Vaned (or closed-vaned) feather*: Pennaceous feathers with barbs held together by barbules and hooklets

References

Benton, M.J. (2004). Vertebrate Paleontology, 3rd Edition. Oxford: Blackwell Publishing.

Buhler, P., Matin, L.D. and Witmer, L.M. (1988). "Cranial kinesis in the Late Cretaceous birds *Hesperornis* and *Parahesperornis*." The Auk, 105: 111-122.

Cambra-Moo, O, Buscalioni, A.D., Cubo, J., Castanet, J., Loth, M.-M., deMargerie, E., and de Ricqlès, A. (2006). "Histological observations of Enantiornithine bone (Saurischia, Aves) from the Lower Cretaceous of Las Hoyas (Spain)." Comptes Rendus Palevol, 5(5): 685-691.

Cau, A. (2011). "L'enigmatico (o forse, no) *Mystiornis*." Theropoda (Weblog entry), 20-May-2011. Accessed online 28-Feb-2012 at <http://theropoda.blogspot.com/2011/05/lenigmatico-o-forse-no-mystiornis.html>

Cau, A. (2012). "*Schizooura*!" Theropoda (Weblog entry), 12-Feb-2012. Accessed online 2-May-2012 at http://theropoda.blogspot.com/2012/02/schizooura.html

Cau, A. and Arduini, P. (2008). "*Enantiophoenix electrophyla* gen. et. sp. nov. (Aves, Enantiornithes) from the Upper Cretaceus (Cenomanian) of Lebanon and its phylogenetic relationships." Atti della Societa Italiana di Scienze Naturali e del Museo ivico di Storia Naturale in Milano, 149(2): 293-324.

Chiappe, L.M. (1993). "Enantiornithine (Aves) tarsometatarsi from the Cretaceous Lecho Formation of Northwestern Argentina." American Museum Novitates, 3083: 1–27.

Close, R.A. and Rayfield, E.J. (2012). "Functional Morphometric Analysis of the Furcula in Mesozoic Birds." PLoS ONE 7(5): e36664.

Dyke, G., Vremir, M., Kaiser, G., and Naish, D. (2012). "A drowned Mesozoic bird breeding colony. from the Late Cretaceous of Transylvania" Naturwissenschaften, 99(6): 435-442.

Gao C., Morschhauser, E.M., Verricchio, D.J., Liu J. and Zhao B. (2012). "A second Soundly Sleeping Dragon: New anatomical details of the Chinese troodontid *Mei long* with implications for phylogeny and taphonomy." PLoS ONE 7(9): e45203.

Habib, M., Hall, J., Hone, D. and Chiappe, L. (2012). "Aerodynamics of the tail in *Microraptor* and the evolution of theropod flight control." 72nd Annual Meeting of the Society of Vertebrate Paleontology, 20 October 2012.

Hall, J., Habib, M., Hone, D. and Chiappe, L. (2012). "A new model for hindwing function in the four-winged theropod dinosaur *Microraptor gui*." 72nd Annual Meeting of the Society of Vertebrate Paleontology, 20 October 2012.

Hill, G.E. (2010). National Geographic Bird Coloration. National Geographic Books.

Holtz, T.R. Jr. (2012). Dinosaurs: The Most Complete, Up-to-Date Encyclopedia for Dinosaur Lovers of All Ages. Winter 2011 Appendix. Accessed online 28-Feb-2012 at <http://www.geol.umd.edu/~tholtz/dinoappendix/HoltzappendixWinter2011.pdf>

Ji Q., Currie, P.J., Norell, M.A., and Ji S. (1998). "Two feathered dinosaurs from northeastern China." Nature, 393(6687): 753-761.

Johnsgard, P. (1987). "Diving Birds of North America: 2 Comparative Distributions and Structural Adaptation. " Papers in the Biological Sciences.

Livezey, B.C. and Zusi, R.L. (2007). Higher-order phylogeny of modern birds (Theropoda, Aves: Neornithes) based on comparative anatomy. II. Analysis and discussion. Zoological Journal of the Linnean Society, 149(1): 1-95.

Longrich, N., Curriw, P.J. and Dong Z.-M. (2010). "A new oviraptorid (Dinosauria: Theropoda) from the Upper Cretaceous of Bayan Mandahu, Inner Mongolia." Palaeontology, 53(5): 945-960.

Longrich, N.R., Vinther, J., Meng Q., Li Q., and Russell, A.P. (2012). "Primitive wing arrangement in *Archaeopteryx lithographica* and *Anchiornis huxleyi*." Current Biology, published online before print 21 November 2012.

Martin, L.D., Kurochkin, E.N., and Tokaryk, T.T. (2012). "A new evolutionary lineage of diving birds from the Late Cretaceous of North America and Asia." Palaeoworld.

Mortimer, M. (2010). The Theropod Database. Accessed online 28-Feb-2012 at <http://home.comcast.net/~eoraptor/>

O'Connor, J.K. (2010). "A revised look at *Liaoningornis longidigitrus* (Aves)." Vertebrata PalAsiatica, 50(1): 25-37.

O'Connor, J.K. and Zhou Z. (2012). "A redescription of *Chaoyangia beishanensis* (Aves) and a comprehensive phylogeny of Mesozoic birds." Journal of Systematic Palaeontology, iFirst 2012, 1-18.

O'Connor, J.K., Chiappe, L.M., Chuong C., Bottjer, D.J., and You H. (2012). "Homology and potential cellular and molecular mechanisms for the development of unique feather morphologies in early birds." Geosciences, 2: 157-177.

Ohmes, D. (2012). "*Microraptor hanqingi*, new species from China." Message to the Dinosaur Mailing List, 25-May-2012. Accessed online 11-Jun-2012 at <http://dml.

cmnh.org/2012May/msg00273.html>

Parsons, W.L. and Parsons, K.M. (2009). "Further descriptions of the osteology of *Deinonychus antirrhopus* (Saurischia, Theropoda)." Bulletin of the Buffalo Society of Natural Sciences, 38: 43-54.

Paul, G.S. (2010). The Princeton Field Guide to Dinosaurs. Princeton University Press.

Prum, R.O. (1999). "Development and early origin of feathers." Journal of Experimental Zoology (Mol Dev Evol), 285: 291-306.

Reynaud, F.N. (2005). "Functional morphology of the hindlimbs of *Hesperornis regalis*: A comparison with modern diving birds." Geological Society of America, 37(7): 133A.

Sanchez, J. (2012). "Diving Birds in the Prairies: Late Cretaceous Hesperornithiformes." Royal Tyrell Museum Speaker Series 2012. Accessed online 17-Sep-2012 at <http://www.youtube.com/watch?v=MKZBSV8TnN0>

Sanz, J.L. and Ortega, F. (2002). "The birds from Las Hoyas." Science Progress, 35(2): 113-130.

Senter, P. (2006). "Comparison of Forelimb Function Between *Deinonychus* And *Bambiraptor* (Theropoda: Dromaeosauridae)". Journal of Vertebrate Paleontology, 26(4): 897–906.

Senter, P. (2007). "A new look at the phylogeny of Coelurosauria (Dinosauria: Theropoda)." Journal of systematic Palaeontology, 5: 429-463.

Senter, P. (2011). "Using creation science to demonstrate evolution 2: morphological continuity within Dinosauria." Journal of Evolutionary Biology, 24(10): 2197-2216.

Sues, H.D. and Averianov, A. (2004). "Dinosaurs from the Upper Cretaceous (Turonian) of Dzharakuduk, Kyzylkum Desert, Uzbekistan." Journal of Vertebrate Paleontology, 24(3).

Sullivan, C., Hone, D.W.E., Xu,X. and Zhang F. (2010). "The asymmetry of the carpal joint and the evolution of wing folding in maniraptoran theropod dinosaurs." Proceedings of the Royal Society B, 277(1690): 2027–2033.

Turner, A.H., Makovicky, P.J., and Norell, M.A. (2012). "A review of dromaeosaurid systematics and paravian phylogeny." Bulletin of the American Museum of Natural History, 371: 206pp.

Walker, C.A., Buffetaut, E., and Dyke, G.J. (2007). "Large euenantiornithine birds from the Cretaceous of southern France, North America and Argentina." Geological Magazine, 144(6): 977-986.

Zelenitsky, D.K., Therrien, F., Erickson, G.M., Debuhr, C.L., Kobayashi Y., Eberth, D.A., and Hadfield, F. (2012). "Feathered non-avian dinosaurs from North America provide insight into wing origins." Science, 338 (6106): 510.

Zheng, F. and Zhou, Z. (2004). "Palaeontology: Leg feathers in an Early Cretaceous bird." Nature, 431: 925.

Zinoviev, A. (2010). "Notes on the hindlimb myology and syndesmology of the Mesozoic toothed bird *Hesperornis regalis* (Aves: Hesperornithiformes)." Journal of Systematic Paleontology, 9(1): 65-84.

Zongker, D. (2007). "Chicken Chicken Chicken." Annals of Improbable Research, 12(6).

Matthew P. Martyniuk is an illustrator and science educator specializing in Mesozoic birds and avian evolution. He has been drawing prehistoric flora and fauna since he first held a pencil, and became fascinated with the dinosaur/bird transition after discovering a copy of Gregory S. Paul's *Predatory Dinosaurs of the World* at his local library. His illustrations and diagrams have appeared in a variety of books, news articles, and television programs from Discovery, the Smithsonian, and the BBC, and he publishes the paleontological blog *DinoGoss*. He is a founding member of "Wikiproject Dinosaurs", an initiative to generate and curate scientifically precise content for the online encyclopedia *Wikipedia*. Additional art and information can be found at his Web site, www.henteeth.com.

PANAVES

Made in the USA
Lexington, KY
06 November 2017